# SpringerBriefs in Biotech Patents

*Series Editor*

Ulrich Storz, Duesseldorf, Germany

For further volumes:
http://www.springer.com/series/10239

Ulrich Storz · Wolfgang Flasche
Johanna Driehaus

# Intellectual Property Issues

## Therapeutics, Vaccines and Molecular Diagnostics

 Springer

Ulrich Storz
Patent Attorneys
Michalski Huettermann & Partner
Duesseldorf
Germany

Johanna Driehaus
Viering, Jentschura & Partner
Duesseldorf
Germany

Wolfgang Flasche
Immatics Biotechnologies GmbH
Martinsried
Germany

ISSN 2192-9904
ISBN 978-3-642-29525-6
DOI 10.1007/978-3-642-29526-3
Springer Heidelberg New York Dordrecht London

ISSN 2192-9912 (electronic)
ISBN 978-3-642-29526-3 (eBook)

Library of Congress Control Number: 2012937289

Printed on acid-free paper

Springer is part of Springer Science+Business Media (www.springer.com)

# Preface

This is the second volume of SpringerBriefs on Biotech Patents. Other than the first volume, which addressed some fundamental issues of Biotech Patents, this volume will deal with three commercially important technical disciplines within the Biotech arena, namely therapeutic antibodies, molecular diagnostics and peptide vaccines and peptide therapeutics.

Therapeutic antibodies are the fastest growing group of protein therapeutics. Antibody patents are subject to similar principles as patents related to small molecular drugs, although some differences apply, particularly with respect to inventive step. The basics principles for protecting antibody compounds will be discussed. Further, additional ways to create follow-up protection for antibody therapeutics will be discussed.

Another important market relates to Molecular Diagnostics. According to the latest figures the molecular diagnostic market in the US alone is worth about \$2.9 billion, with a predicted annual growth of 15% until 2015, resulting in a volume of \$6.2 billion. This alone justifies a closer look at the relevant patent issues accompanying these developments. An overview of the patent landscape in molecular diagnostics is thus provided, and issues of patentability with respect to the different technologies and compounds used therein are discussed.

Peptide Vaccines and Peptide Therapeutics are increasingly entering into the focus of pharmaceutical companies. However, peptide patents pose particular problems which are quite uncommon even for skilled biopatent experts. The peptide section is intended to give a broad overview of areas of law that are particularly relevant to the patenting of peptide vaccines and therapeutic peptides as products and in compositions. The scope of patentable subject matter will be discussed, as it has been the focus of much wrangling and debate in the courts. Further, a comparative look is provided at how American and Chinese authorities treat peptide-based inventions.

We hope that this volume will be helpful for patent practitioners to be able to anticipate, or appropriately respond to, problems coming up in the patent prosecution process, and to develop a suitable patent strategy with respect to these capital intensive technologies.

Duesseldorf

Ulrich Storz
Wolfgang Flasche
Johanna Driehaus

# Contents

# IP Issues of Therapeutic Antibodies

Ulrich Storz

**Abstract** The high investments necessary to bring an antibody therapeutic to the market require a sound patent strategy. Although compound protection provides the broadest scope of protection, other ways of follow-up protection should be considered by innovators to achieve as long protection as possible. Further, in case a theoretical antibody against a given target is already prior art, innovators should be aware of methods to create compound protection for second or higher generation antibodies.

**Keywords** Patent · IP · Antibodies · Compound · Protection · Biosimilars

## 1 Introduction

Therapeutic antibodies are the fastest growing group of protein therapeutics. With a limited set of underlying technologies, drugs for a wide area of indications, including cancer, autoimmunity, neurodegeneration, and infections, can be generated.

Table 1 shows the therapeutical antibodies which have, in 2010, achieved global sales of one billion USD or more (data taken from company information).[1]

---

[1] *Note* The phrase "key patent" refers to only one member of a patent family that exists for the product. INN, international non-proprietary name.

---

U. Storz (✉)
Michalski Huettermann & Partner Patent Attorneys, Neuer Zollhof 2,
40221 Duesseldorf, Germany
e-mail: st@mhpatent.de

U. Storz et al., *Intellectual Property Issues*, SpringerBriefs in Biotech Patents,
DOI: 10.1007/978-3-642-29526-3_1, © The Author(s) 2012

**Table 1** Therapeutic mAbs which have, in 2010, achieved global sales ≥1 billion USD

| Antibody | Brand name | Company | Key indication | Target | Key IP right US | Key IP right EP | Gobal sales in 2010 (billion USD) |
|---|---|---|---|---|---|---|---|
| Infliximab | Remicade | J&J | Rheumatoid Arthritis | TNFα | | EP0666868 | 8.0 |
| Bevacizumab | Avastin | Roche | Colon cancer | VEGF-A | US7060269 | EP1167384 EP1325932 | 6.8 |
| Rituximab | Rituxan | Roche | Non Hodgkin Lmyphoma | CD20 | US5736137 | EP0669836 | 6.7 |
| Adalimumab | Humira | Abbott | Rheumatoid Arthritis | TNFα | US6090382 US6509015 | EP0929578 | 6.5 |
| Trastuzumab | Herceptin | Roche | Breast Cancer | HER2/neu | US6719971 | EP0590058 | 5.5 |
| Cetuximab | Erbitux | BMS | Colon, head, and neck cancer | EGFR | US6217866 | EP0359282 | 3.2 |
| Ranibizumab | Lucentis | Novartis Roche | Wet Macular degeneration | VEGF-A | US6407213 | EP0940468 | 3.1 |
| Natalizumab | Tysabri | Biogen Idec | Multiple sclerosis | $\alpha^4$-integrin | US5840299 | EP0804237 | 1.75 |
| Omalizumab | Xolair | Roche Novartis | Allergic Asthma | IgE Fc | US6267958 | EP0841946 | 1.1 |
| Palivizumab | Synagis | Astra Zeneca | RSV infection | RSV protein F | US5824307 | EP0783525 | 1.0 |

As discussed earlier in this book series,[2] the development of a new drug is a costly and time-consuming matter. According to a study performed by the Tufts Center for the Study of Drug Development, the estimated average costs of developing a new Biologic are 1.2 billion USD,[3] while development times are slightly longer than those reported for small molecule drugs.[4]

In order to recover the expenses invested into research and development of new biopharmaceutics, particularly into a new antibody, patents are an indispensable tool, as they provide an exclusive right with respect to the protected subject matter. Only the patentee, or his licensee, is thus allowed to exploit the invention commercially, e.g., by marketing the protected antibody.

Antibodies are proteins and, as such, chemical compounds. For this reason, antibody patents are subject to similar principles as patents related to small molecular drugs, although some differences apply, particularly with respect to inventive step.[5] The basic principles for protecting antibody compounds will be discussed in the following. Further, additional ways to create follow-up protection for antibody therapeutics are discussed.

## 2 Compound Protection

Compound protection is probably the most important protection antibody companies can rely on, as it provides an exclusive right to offer and sell the respective antibody on different markets. Furthermore, while patents protecting a particular technology expire after, roughly, two decades, it remains still possible to achieve compound protection for a new antibody even after expiry of the respective method patents used to generate, or produce, the said new antibody.

Different possibilities exist to specify an antibody in a patent. As a general rule, the specification of an antibody by its target provides the broadest scope of protection, but can only be achieved at a very early stage, e.g., when the target has been discovered and "deorphaned", i.e., attributed with a physiological role. A patent protecting second or higher generation antibodies against the said target has, usually, a narrower scope, because more technical details (e.g., structural or functional) are required in the patent claim to make it novel over the prior art. In even later stages, combinations comprising an antibody against the said target plus one or more other compounds, or dosage regimen or new medical uses of the said antibody, can be claimed, yet patents of this type have an even narrower scope of

---

[2] Storz (2010).

[3] Average Cost to Develop a New Biotechnology Product Is $1.2 Billion. Tufts Center for the Study of Drug Development, November 9, 2006. Available at http://csdd.tufts.edu/NewsEvents/NewsArticle.asp?newsid=69. Accessed on October 27, 2009.

[4] Grabowski et al. (2006).

[5] Stewart et al. (2011).

protection, allowing, e.g., off-label use by personal prescription. In the following, the different types of protection will be discussed briefly. In Table 2, examples are given for each type.

## 2.1 Specification by Target

In case a new protein has been discovered and a therapeutic use thereof has been disclosed, both the European Patent Office (EPO) and the United States Patent and Trademark Office (USPTO) grant claims related to a theoretical antibody against the said protein,[6,7] even if the applicant has not produced a real antibody, or provides no data or enablement related to such antibody. Claims of such type have, obviously, the broadest scope, as they encompass all future antibodies against the said target put into practice later on (i.e., during the 20 years after filing of the target patent). The offices' rationale is that the provision of a novel protein X enables a skilled artisan to produce an antibody against said protein. Therefore, it is considered a fair reward for the applicant of protein X to be granted a claim related to an antibody against the protein. Once granted, the scope of protection of such claim extends to next generation antibodies against protein X as well. This means that somebody who provides a well-defined antibody against protein X will be, in his right to practice, dependent on the assignee of the first-generation patent, despite the fact that the said assignee has never provided a "real" antibody, and although he himself might as well be awarded a Patent on his antibody.

While cellular signaling processes are today well understood, new potential targets for antibody therapy are still being discovered. Today, about 100 such targets are addressed by approved biopharmaceuticals,[8] but the spectrum of soluble proteins or membrane receptors yet undiscovered that represent potential therapeutic targets should be much higher. Although the evaluation of a new target and the subsequent development of a respective antibody are costly endeavors, recent advancements in antibody technology may accelerate the validation of new targets, in particular those relevant to cancer, autoimmune diseases, infectious diseases, and neurodegenerative diseases. Patents which merely claim any conceivable antibody against a new target will thus be a frequent sight even in the future.

However, with respect to the currently established targets (see Table 1), the respective patents are about to expire, or have already expired. In case a company develops a second antibody against a target addressed by these antibodies, other strategies are necessary to protect such second generation antibodies.

---

[6] EPO decision T542/95.

[7] Noelle v. Lederman, 355 F.3d 1343, 2004 U.S. App. LEXIS 774.

[8] Overington et al. (2006).

**Table 2** Examples for wording of antibody compound claims

| Specification by | Example | Claim wording |
| --- | --- | --- |
| Target | US5654407 (Bayer) | 1. A composition comprising human monoclonal antibodies that bind specifically to human tumor necrosis factor alpha. |
| Target-independent function | US7214775 B2 (Biowa) | 1. An antibody composition comprising antibody molecules, wherein 100% of the antibody molecules comprising a Fc region comprising complex N-glycoside-linked sugar chains bound to the Fc region through N-acetylglucosamine of the reducing terminal of the sugar chains do not contain sugar chains with a fucose bound to the N-acetylglucosamines. |
| Epitope | US8080247 (Janssen) | 1. An isolated human anti-IL-12 antibody, wherein said antibody binds to a conformational epitope of the IL-12 protein comprising residues 15, 17–21, 23, 40–43, 45–47, 54–56, and 58–62 of the amino acid sequence of SEQ ID NO:9. |
| Sequence CDR | EP1309691 (J&J) | 1. An antibody comprising the heavy chain complementarity determining regions (CDRs) and variable framework regions (FRs) of mAb TNV148 as described in Fig. 4; and the light chain CDRs and variable FRs of mAb TNV 148 as described in Fig. 5; optionally further comprising the specified substitution from proline to serine in FR3 of mAb TNV148B as described in Fig. 4. |
| Sequence VL and VH | EP0590058 (Genentech) | 3. A humanized Antibody which comprises a VL domain comprising the polypeptide sequence DIQMTQSPSSLSASVGDRVTITCRASQ DVNTAVAWYQQKPGKAPKLLIYSASF LESGVPSRFSGSRSGTDFTLTISSLQPE DFATYYCQQHYTTPPTFG QGTKVEIKRT and a VH domain comprising the polypeptide sequence EVQLVESGGGLVQPGGSLRLSCAASGFNIK DTYIHWVRQAPGKGLEWVARIYPTNGY TRYADSVKGRFTISADTSKNTAY LQMNSLRAEDTAVYYCSRWGGDG FYAMDVWGQGTLVTVSS. |
| Target affinity | US6090382 (Abbot) | 1. An isolated human antibody, or an antigen binding portion thereof that dissociates from human TNFα with a $K_d$ of $1 \times 10^{-8}$ M or less and a $K_{off}$ rate constant of $1 \times 10^{-3}$ $s^{-1}$ or less, both determined by surface plasmon resonance, and neutralizes human TNFα cytotoxicity in a standard in vitro $L_{929}$ assay with an $IC_{50}$ of $1 \times 10^{-7}$ or less |
| Competitive binding | US7595378 (Genmab) | 4. An isolated human monoclonal antibody which binds to human EGFR, competes with antibody 225, but does not compete with antibody 528. |
| Deposited cell | US6582959 (Genentech) | A monoclonal antibody produced by the hybridoma cell deposited under American Type Culture Collection Accession Number ATCC HB10709. |
| Product-by-process | RE32011 (Scripps) | 8. An antibody which catalyzes hydrolysis of beta -amyloid at Val39–Val40, Phe19–Phe20 or Phe20-Ala21 of SEQ ID NO: 1, the antibody being produced by a method comprising immunizing an animal with a transition state analog which mimics the transition state that beta-amyloid adopts during hydrolysis, the transition state analog being selected from a group consisting of statine and phenylalanine-statine. |

## 2.2 Specification by Target-Independent Functional Properties

Antibodies can have functional properties which are target-independent. The development of such a new functional property can thus give rise to a patent the scope of which extends to all antibodies, irrespective of the target they bind, which have such property. This can, for example, apply to increased effector function by sequence engineering (e.g., US7863419 by Biogen) or post-translational glycoengineering of the Fc region (e.g., US7214775 by Biowa, or EP1071700 by Glycart), to increased serum half-life by Fc glycoengineering (e.g., US7361740 by Protein Design Labs), or to increased antibody-dependent cell cytotoxicity (ADCC) by using a novel expression system which, as such, creates an N-glycan structure which is essentially fucose-free (e.g., WO2011107520 by Cilian AG).

## 2.3 Specification by Epitope

Another way to create patent protection for a second generation antibody is to claim a specifity against a given epitope, or subdomain, of a target, in case the said epitope has not yet been described as clinically relevant. This approach makes sense in case the target as such has already been described, particularly when blocking only a specific epitope instead of the whole target may yield at least a theoretical benefit. However, the scope of a patent claiming an antibody against a specific epitope of a given target will not encompass antibodies against other epitopes of the same target.

This has been, for example, decided by a US District Court[9] recently in a case where Genentech and Biogen sued GlaxoSmithKline (GSK) and Genmab for infringement of a patent protecting Rituxan, namely by GSK's Arzerra. Both Arzerra and Rituxan target CD20; however, Arzerra binds an epitope of the latter different from Rituxan, and with a different affinity. US Patent 7682612 claims the treatment of Chronic Lymphatic Leukemia by administration of an anti-CD20 antibody, and is thus not per se restricted to a particular epitope thereof. However, in order to overcome an office objection related to lack of enablement, Biogen has, during the patent prosecution, stated that the term "anti-CD20 antibody" shall mean antibodies having similar affinity and specifity as Rituxan. The respective court construed the patent claims as being restricted to anti-CD20 antibodies having similar affinity and specifity as Rituxan and thus concluded that Arzerra would not fall under the scope of the said patent. Although the claim language as such was not restricted to Rituxan, the court construed the claim in such way because of the applicant's statements made in the prosecution history.

---

[9] U.S. District Court for the Southern District of California; case 10-CV-00608 BEN (GS) of Oct 17, 2011.

## 2.4 Specification by Target-Dependent Functional Properties

Another way to create patent protection for a second generation antibody is to specify the latter through target-dependent functional properties, e.g., binding affinity against a given target, or competitive binding. The former is often done by claiming a minimum affinity to a target. In such case, all later antibodies having even better affinity will fall under the scope of protection of such patents, even if they have no material relationship to the antibody which has been provided by the patentee. Existing patents with these claims are a real threat to competitors, particularly to those specializing in antibody optimization ("Biobetters").

## 2.5 Specification by Sequence

Yet another way to create patent protection for a second or higher generation antibody is to specify a sequence thereof. Claimed sequences are commonly specified in such a way that, besides the mere sequence, a certain similarity interval (e.g., 85%) is comprised as well. In antibody claims, this makes little sense as the specifity of a given antibody is highly dependent on its sequence. Therefore, second or higher generation antibody claims are commonly drafted in such form that a DNA or AA sequence is claimed (e.g., SEQ ID No 1), sometimes together with possible variations (e.g., R112T). The scope of protection is thus clearly defined, yet quite narrow. Competitors which replace one of the claimed residues by a residue which is not claimed thus no longer fall under the literal scope of the patent, although the antibody may retain its function despite such modification.

Most legal systems provide doctrines of equivalents to anticipate situations as discussed above. As a rule of thumb, German judges[10] tend to provide a broader scope of equivalence than UK judges,[11] although attempts have been made under the European Patent Convention to establish a uniformal definition of the term "equivalent".[12] There is, however, no court decision in the US or in Europe which has yet defined the scope of equivalence for biosequence claims. This means that it is uncertain how far a competitor must amend the claimed sequence to make sure not to be sued for equivalent infringement.

It is yet noteworthy that, after the "Festo" decision issued by the U.S. Supreme Court,[13] legal action related to equivalent infringements can no longer be enforced

---

[10] Three-step approach, as applied in the BGH decisions "Kunststoffrohrteil", "Schneidmesser I", "Schneidmesser II", "Custodiol I", "Custodiol II", GRUR 2002, 511–531.

[11] "Catnic test" as applied in Kirin-Amgen, Inc. v. Hoechst Marion Roussel Ltd. (2004) UKHL 46 (2004-10-21).

[12] Article 2 of the Protocol on the Interpretation of Article 69 EPC.

[13] Festo Corp. versus Shoketsu Kinzoku Kogyo Kabushiki Co., 535 U.S. 722 (2002).

in the US if, during patent prosecution, the scope of the patent has been narrowed in such a way that the alleged infringement is no longer covered by the literal scope of protection (so-called "prosecution history estoppel").

The effect of this ruling on antibody sequence claims which are narrowed down during prosecution (e.g., from a sequence claim reciting "amino acid Seq. ID No. 1 and any sequence which has 85% identity to the former" to a claim which is restricted to the mere Seq. ID No. 1) has not made its way into case law yet, but it is to be expected that, in such case, competitors can easily circumvent the scope of protection by amending a single amino acid residue only. This requires that applicants draft their patent claims with caution, while competitors should always have a look at the patent prosecution history.

Because of the legal uncertainties with respect to the equivalence problem, applicants should restrict the length of the claimed sequence to the bare minimum needed to meet the novelty/inventive step requirement. From that perspective, it is preferable to only specify e.g. one or more CDRs in the claims, instead of a full variable domain, let alone a full heavy or light chain, because for a competitor it is much more difficult to replace one or more amino acids in the CDRs than, say, in the Fc region, at least in case he does not want to compromise the binding behavior of his antibody.

However, even in case an antibody claim has been narrowed down, during prosecution, to the exact sequence of either a CDR, or even of heavy and/or light chain, and given that the mere substitution of one amino acid residue in the claimed sequence would isolate a competitor from patent infringement because of an unexpectedly restrictive claim interpretation by the courts with virtually no scope of equivalence, fallback positions still exist for innovators.

In that case, a patentee could still take benefit from the strict rules the European Medicines Agency (EMA) applies for follow-on biologics (also called biosimilars)[14] for which an accelerated approval under the biosimilar pathway[15] is requested. In order to qualify as a biosimilar, the EMA requires that the follow-on biologic has exactly the same amino acid sequence as the innovator biologic.[16] In case a competitor modifies the amino acid sequence of a given innovator biologic in order to avoid patent infringement, it is quite unlikely that his follow-on version would qualify for accelerated approval, at least in Europe. This means that the said competitor would have to go through full scale approval to obtain market authorization— quite an unattractive option for most biosimilar manufacturers.

---

[14] Biosimilars will be discussed in a later volume of this book series.

[15] See, among others, guideline EMEA/CHMP/42832/2005.

[16] The EMEA defines Biosimilars as follows: "The active substance of a similar biological medicinal product must be similar, in molecular and biological terms, to the active substance of the reference medicinal product. For example, a medicinal product containing interferon alfa-2a (…) should refer to a reference medicinal product containing as its active substance interferon alfa-2a. Therefore, a medicinal product containing interferon alfa-2b could not be considered as the reference medicinal product".

## 2.6 Specification by Deposited Cell Line or Product-by-Process

The deposition of a cell line, e.g., a hybridoma cell line or a transfected host cell line, may be an adequate way of specifying an antibody in order to avoid sequencing errors and typographical errors, or to provide enabling information for features which relate to post-translational modifications (e.g., unusual glycosylation patterns). The deposition process is subject to laws and bylaws provided by the respective Patent legislations, while the respective claim language simply refers to the deposition nomenclature of the deposited cell line. Product-by-process claims define a compound by the way it has been produced, but still remain veritable compound claims. The claim language is, for example, "Antibody obtained by process X". While before the EPO, such claim type is only allowable if the antibody cannot be defined in a sufficient manner on its own (e.g., by binding properties or sequence), there seems to be no such restrictions in the US.

It is important that both Deposited Cell Line claims and Product-by-Process claims provide full compound protection irrespective of the actual method used (which means that the above claim is actually read as "Antibody obtainable by process X"), at least in Europe,[17] whereas in the US the case law is inconsistent for Product-by-Process claims. In a case related to a method of preparing Factor VIII,[18] the Court of Appeals for the Federal Circuit held that product-by-process claims were to be construed as product claims, independent of how the product was made, while in the younger case another panel of the same court held that such claims are limited to a product that is prepared by the specific process steps recited therein.[19] Therefore, it appears that in the US, product-by-process claims are merely disguised process claims, in that they only protect a claimed compound in case it has actually been produced with the specified method.

## 3 Other Types of Follow-Up Protection

Once compound protection of a given antibody has expired, other types of protection are still available which can be used to extend the term of protection for the said antibody. In most cases, however, the scope of protection which can be achieved is smaller than with compound protection. In the following, the most important options for follow-up protection will be discussed. A summary is given in Table 3.

---

[17] Scripps Clinic & Research Foundation v. Genentech, Inc., 18 USPQ 2nd 1001 (Fed. Cir. 1991).

[18] Scripps Clinic & Research Foundation v. Genentech, Inc., 18 USPQ 2nd 1001 (Fed. Cir. 1991).

[19] Abbott Labs. v. Sandoz, Inc., 566 F.3d 1282, 1300 (Fed. Cir. 2009).

## 3.1 Second Medical Use

In some cases, a novel indication (second medical use) for a known antibody is discovered at a later stage (as it was the case with Avastin, which turned out to be effective also for the treatment of age-related macular degeneration). Such novel indication can give rise to a second or higher generation patent, too. The respective patent claim for such use will be as follows: "Antibody XY for the treatment of disease Z". It is to be noted that in Europe such wording does not qualify as a method of treatment (which is not patentable under EPC), after the EPC has been revised in 2007,[20] thus making the Swiss-type claim wording obsolete.[21]

In the US, such claim wording is not accepted, as "use" is not a claim category provided by the U.S. Patent Act.[22] Therefore, the claim wording should be as follows: "A process comprising administering a composition comprising Antibody XY to a human in an amount effective for treating a disease Z".

Applicants must however be aware that, in case the compound protection for a given antibody has expired, they cannot avoid offlabel use of a biosimilar of the said antibody for the protected second medical indication upon individual prescription. However, patentees can block a competitor from advertising and even labeling his follow-on biological with respect to the said indication. Further, particular constellations exist in which the said competitor can be sued for indirect infringement.

## 3.2 Combination Therapy

In some cases, the use of an antibody together with another agent (e.g., another antibody, a chemotherapeutic drug, or the like) turns out to have beneficial or even synergistic effects (see for example the combination of Methotrexate and anti-TNFα Antibody for the treatment of rheumatoid arthritis, as claimed in Abbott's US7223394). The respective claim wording for such a combination could for example read as follows: "Use of Antibody XY in combination with agent Z", or "Composition comprising Antibody XY and agent Z for the treatment of disease Z". Other patents related to combination therapies are ImClone's US6811779 (combination of VEGF antibody and radiation), Yeda's US6217866 (combination

---

[20] Article 54(5) EPC.

[21] The Swiss-type claim wording ("Use of a substance or composition X for the manufacture of a medicament for treatment of disease or condition Y") had been found acceptable by the enlarged board of appeal in decision G 5/83 avoid a violation of the exclusion of therapeutic methods under old EPC (Article 52 (4) EPC 1973).

[22] 35 U.S.C. 101: "Whoever invents or discovers any new and useful process, machine, manufacture, or composition of matter, or any new and useful improvement thereof, may obtain a Patent therefore".

**Table 3** Other possibilities for second or higher generation of antibody patents

| Type | Example | Claim language |
|---|---|---|
| Second medical use | EP1616572 (Genentech) | 1. Use of rituximab in the manufacture of a medicament for treatment of chronic lymphocytic leukemia (CLL) in a human patient, wherein the medicament is for administration to the human patient at a first dose of 375 mg/m$^2$ and subsequent dosage of 500–1500 mg/m$^2$ |
| Combination therapy | EP1169059 (Aventis) | 1. A pharmaceutical combination, comprising Docetaxel and rhuMAb HER2 |
| Dosage regimen | EP1616572 (Genentech) | 1. Use of rituximab in the manufacture of a medicament for treatment of chronic lymphocytic leukemia (CLL) in a human patient, wherein the medicament is for administration to the human patient at a first dose of 375 mg/m$^2$ and subsequent dosage of 500–1500 mg/m$^2$ |
| Formulation | EP1687031 (Merck) | 1. Aqueous composition consisting of a solvent which consists at least partly of water, an anti-EGFR (epidermal growth factor receptor) antibody, a buffer, an amino acid, a surfactant, and sodium chloride as isotonic agent |

of EGFR antibody and chemotherapy, and Aventis' EP1169059 (pharmaceutical combination, comprising Docetaxel, and rhuMAb HER2).

Patents claiming combination therapies will become particularly important with the introduction of personalized medicine, where particular drug combinations may be selected to treat a patient with a given genetic predisposition. Further, similar issues apply with respect to offlabel use and indirect infringement as discussed above with second medical use.

## 3.3 Dosage Regimen

In 2010, the Enlarged Board of Appeal of the EPO made clear that even a new dosage regimen of a known drug or drug combination can be subject of a patent if the said regiment has surprising, or beneficial, properties.[23] The respective claim wording for such combination could, for example, read as follows: "Use of compound A for the treatment of disease X by administration once per day prior to sleep", or "Use of compound A for treatment of disease X, wherein the medicament is administered to the patient at a first dosage Y and subsequent dosage Y."

Because such claim type is relatively new, it is difficult, to date, to predict what kind of evidence an applicant has to provide to meet the inventive step requirement for such claim.

Further, similar issues apply with respect to offlabel use and indirect infringement as discussed above with second medical use.

---

[23] Decision G2/08.

## 3.4 Formulation and Galenics

In the small molecule pharma market, the filing of second or higher generation patents protecting a new formulation or galenic of a drug is well-established practice to extend the protection for a given drug, and has been accepted by the courts, too.[24] Formulations play an even more important role in antibody drugs, because the latter are extraordinarily large molecules (IgG have about 150 kD) which can be the subject of aggregation and artifacts occurring during storage. Further, a new formulation can give rise to a new way of administration of a given antibody. Because formulation claims and many galenic claims can be drafted as compound claims ("Formulation comprising antibody X, buffer Y and agent Z") they cannot be bypassed by off-label use, although third parties can formulate the said antibody in a different way in case the compound protection for the said antibody has expired.

# 4 New Antibody Formats

Protein therapeutics which are derived from the general IgG concept are commonly called "new antibody formats". Established formats such as chimerized antibodies, (e.g., US5585089 assigned to Protein Design Labs), humanized antibodies (e.g., US5859205 assigned to Celltech), antigen binding fragments (Fab), single chain variable fragments (scFv, e.g., US5260203 assigned to Enzon), and Receptor-Fc fusion peptides (e.g., US5610279 assigned to Roche for Enbrel) were, strictly speaking, considered "new antibody formats" when first introduced. The basic concept of rearranging and recombining different components of IgGs was further pursued in the last decade.

The potential advantages of new antibody formats compared with full-size molecules depend on the respective nature of the format, and encompass, for example, lack of glycosylation, lack of disulfide bridges, reduced molecular weight, better stability and serum half-life, better tissue penetration, lower immunogenicity, straightforward transfer from animal trials to humans, suitability for oral administration, expression advantages (e.g., expression in *E. coli* or yeast instead of Chinese hamster ovary cells), higher expression efficiency, and ease of selection/screening. These advantages can be referred to meet the inventive step/non-obviousness requirement in patents claiming the respective antibody formats or their technology. Some patents protecting major advancements in this field are listed in Table 4.

---

[24] Unigene Labs., Inc. v. Apotex, Inc. 06-CV-5571 (Fed. Cir. 2011).

**Table 4** Examples for patents protecting new antibody formats

| Company | Technology | Technology name/ candidate drug | Key IP right |
|---|---|---|---|
| Enzon | Polyalkylene oxide-modified scFv | | US7150872 |
| Macrogenics | Diabodies | | US2007004909 |
| CAT | Diabodies (scFv$_2$, potentially bispecific) | | US5837242 |
| Micromet | Bispecific scFv$_2$ directed against target antigen and CD3 on T cells | "BITE" | US7235641 |
| Affimed | Diabody−Diabody dimers | "TandAbs" | US2005089519 |
| Affimed | scFv -Diabody-scFv | "Flexibodies" | US2005079170 |
| Unilever | Camelid antibodies (CH2-CH3-VHH)$_2$ | | US6838254 |
| Ablynx | (camelid VHH) | "Nanobodies" *ATN-103 (anti-TNF)* | US2003088074 |
| Domantis/ GSK | Variable regions of heavy (VH) or light (VL) chain ("Domain Antibodies") | "dAb" | US2006280734 |
| Scancell | Tumor epitopes on a IgG structure with unchanged FC domain | "Immunobody" | US2004146505 |
| Hybritech/ Liliy | Trifunctional antibodies (Fab-Fab-Fab, maleimide linkers) | | US5273743 |
| Trion Pharma | Trifunctional IgG, Fc binds accessory cells, Fabs bind CD3, and tumor antigen | "Triomab" | US6551592 |
| Affitech | Antibodies with T cell epitopes between ß-strands of constant domains, and new V-regions specific for antigen presenting cells | "Troybodies" | US6294654 |
| Affitech | Antibody fragments that cross-link antigen and antibody effector molecules | "Pepbodies" | US2004101905 |
| Vaccibody AS | Bivalent homodimers, each chain consisting of scFv targeting unit specific for antigen presenting cells | "Vaccibody" | US2004253238 |
| Planet Biotechnology | IgA (two IgG structures joined by a J chain and a secretory component), expressed in a plant host, secretory component replaced by a protection protein | "SIgA plAntibodies" | US6303341 |
| Trubion | Variable regions of heavy (VH) and light (VL) chain + Fc (small modular immunopharmaceuticals) | "SMIP" | US2008227958 |
| Haptogen/ Pfizer | Homodimeric heavy chain complex found in immunized nurse sharks, lacking light chains | "NAR" (Novel Antigen Receptor) | US2005043519 |
| AdAlta | Recombinant shark Antibody domain library | "IgNar" | US2009148438 |
| Xencor | Altered Fc region to enhance affinity for Fc gamma receptors, thus enhancing ADCC | "XmAB" | US20080181890 |

(continued)

**Table 4** (continued)

| Company | Technology | Technology name/ candidate drug | Key IP right |
|---------|-----------|--------------------------------|--------------|
| Arana | New world primate framework + non-new world primate CDR | "syn-humanisation" | US2008095767 |
| City of Hope | Dimerized construct comprising CH3 + VL + VH | "minibody" | US5837821 |
| Seattle Genetics | Antibody–drug conjugate technology with enzyme-cleavable linkers | | WO2009117531 |
| Epitomics | Humanized rabbit antibodies with increased target affinity | "RabMAbs" | US2005033031 |
| F-Star | CH2 and CH3 domains with two identical antigen binding sites engineered into the CH3 domains | "Fcab"(antigen binding Fc) | US2009298195 |
| | IgG with two additional binding sites engineered into the CH3 domains | "mAb$^2$" | US2009298195 |
| Symphogen | Polyclonal antibody mixtures obtained by simultaneous expression; antibodies bind to different regions of the same antigen or multiple antigens | "Sympress"*Sym004 (anti EGFR)* | EP2152872 |
| Genmab | IgG4 antibodies with hinge region removed (no interaction with immune system) | "UniBody" | WO2010063785 |
| | Human bispecific mAbs | "UniBody" | US2010105874 |
| Regeneron | Fusion peptides consisting of the extracellular domain of protein receptor + Fc domain | *VEGF trap* extracellular segments of VEGFR1 and 2 + Fc, binds VEGF-A + PLGF | US7087411 |
| Roche | | *Enbrel* | US5610279 |
| Philogen | Fusion proteins for targeted delivery of bioactive molecules to vascular sites of disease | "Vascular Targeting"*L19-TNF-α* | US2010316602 |

# 5 Antibody Mimetics

Proteins not belonging to the immunglobulin family, and even non-proteins such as aptamers or synthetic polymers, have also been suggested as alternatives to antibodies.[25] One reason for the increasing interest in these so-called "alternative scaffolds", or "antibody mimetics", is the blocking effect created by existing antibody patents. As with new antibody formats, potential advantages of new

---

[25] Gebauer and Skerra (2009).

**Table 5** Examples for patents protecting new antibody mimetics

| Company | Scaffold protein | Technology name | Size kDa | Example drug | Key IP right |
|---|---|---|---|---|---|
| Molecular Partners | Ankyrin Repeat Proteins | "DARPins" | 10–19 | MP0112 (Anti VEGF) | US7417130 |
| Borean Pharma | C-Type Lectins | "Tetranectins" | | | US2004132094 |
| Affibody | A-domain proteins of *S.aureus* | "Affibodies" | 6 | ABY-025 (anti Her2) | US5831012 |
| BioRexis/Pfizer | Transferrin | "Transbodies" | | | US2004023334 |
| Pieris Proteolab | Lipocalins | "Anticalin" | 20 | PRS-050 | US7250297 |
| Adnexus/Bristol Myers Squibb | 10th type III domain of fibronectin | "AdNectins" (Monobodies) | 10 | Angiocept (anti VEGFR2) | US6818418 |
| Dyax | Kunitz domain protease inhibitors | | 6 | Ecallantide (anti Kallikrein) | US2004209243 |
| Scil Proteins GmbH | Ubiquitin derived binders | "Affilin" | 10 | SPVF2801–30 (anti-EDB) | US7833629 |
| | Gamma Crystallin derived binders | | 20 | | |
| Selecore/Nascacell | Cysteine knots or knottins | "Microbodies" | | | US7186524 |
| General Hospital/Genetics Institute | Thioredoxin A scaffold | "peptide aptamers" | | | US6004746 |
| Archemix | Nucleic acid aptamers | | | Macugen (anti VEGF); ARC1779 (anti vWillebrandt) | US5475096 |
| Catalyst Biosciences | Target specific proteases obtained by directed evolution | "Alterases" | | | US2004146938 |
| Mosbach/Lund University | Artificial antibodies produced by molecular imprinting of polymers | "plastic Antibodies" | | | US2004157209 |
| Phylogica | Peptide libraries from bacterial genomes | "Phylomers" | | | US6994982 |
| NextBiomed | SH-3 domains | | | | US6794144 |
| Gliknik | Antibody mimetics | "Stradobody" | | | US2010239633 |
| Avidia/Amgen | "A domains" of membrane receptors stabilized by disulfide bonds and Ca$^{2+}$ | "Avimers","Maxibodies" | 9–18 | | US7803907 |
| Evogenix/Cephalon | CTLA4-based compounds | "Evibody" | | | US7166697 |
| Covagen | Fyn SH3 | "Fynomers" | 7 | | US2010119446 |

antibody mimetics, which may give rise to sufficient inventive steps, depend on their respective structural characteristics. These specific advantages may be used as a basis for patentability, i.e., in order to meet the requirements toward novelty and inventive step/non-obviousness. Some selected approaches are shown in Table 5. Some product candidates derived from these approaches have already entered the clinical phase, while others are still in the preclinical phase.

# 6 Conclusion

The high investments necessary to bring an antibody therapeutic to the market require a sound patent strategy. Although compound protection provides the broadest scope of protection, other ways of follow-up protection should be considered by innovators to achieve as long protection as possible. Further, in case a theoretical antibody against a given target is already prior art, innovators should be aware of methods to create compound protection for second or higher generation antibodies.

# References

Grabowski H, Cockburn I, Long G (2006) The market for follow-on biologics: how will it evolve? Health Aff 25(5):1291–1301
Gebauer M, Skerra A (2009) Engineered protein scaffolds as next-generation antibody therapeutics. Curr Opin Chem Biol 13:245–255
Overington JP, Al-Lazikani B, Hopkins AL (2006) How many drug targets are there? Nat Rev Drug Discov 5:993–996
Storz U (2012) Patent lifecycle management, supplementary protec tion certificates and data exclusivity in biopharmaceutics. In: Storz (ed) SpringerBriefs in Biotech Patents, pp 1–12. doi:10.1007/978-3-642-24846-7
Stewart M, Kent L, Smith A, Bassinder E (2011) The special inventive step standard for antibodies. EPI Inf 2:72

# Peptide Vaccines and Peptide Therapeutics

Wolfgang Flasche

**Abstract** Peptide vaccines and Peptide therapeutics are increasingly entering into the focus of pharmaceutical companies. This section is intended to give a broad overview of areas of law that are particularly relevant to the patenting of peptide vaccines and therapeutic peptides as products and in compositions. The scope of patentable subject matter will be discussed, as it has been the focus of much wrangling and debate in the courts.

**Keywords** Peptide · Vaccine · Therapeutic · Patent · Novelty · Disclosure · Enablement

## 1 Introduction

Despite the ongoing hype about antibodies and small molecules still being by far the most abundant pharmaceutics the role of pharmaceutical active peptides is growing. Peptides are currently used in therapeutics, vaccines, and diagnostics. While a few years ago the production of peptides with a sufficient purity was a major concern, the problems today are by far more related to the world of patent and regulatory affairs.

Nowadays, even long peptides can be produced chemically or with recombinant means using well established techniques in a purity that was unheard of 20 years ago. Unfortunately, we observe that it is getting harder and harder to get consistent patents issued in different jurisdictions. When dealing with peptides one has to act

W. Flasche (✉)
Immatics biotechnologies GmbH, Fraunhoferstraße 18b, 82152 Martinsried, Germany
e-mail: w.flasche@immatics.com

U. Storz et al., *Intellectual Property Issues*, SpringerBriefs in Biotech Patents,
DOI: 10.1007/978-3-642-29526-3_2, © The Author(s) 2012

in a world that is between DNA/RNA and other bio inventions on the one hand, and chemical inventions such as small molecules, on the other. Issues are, to name a few, unity, sufficient data, and sufficient disclosure.

We will try to highlight the problems connected to peptides in patent law and the application of the law by the patent authorities' TOs from a European, American, and Chinese view.

## 2 Definition of Peptides

Peptides (from the Greek πεπτός, "digested" from πέσσειν "to digest") are short polymers of amino acid monomers linked by peptide bonds also known as amide bonds –CO–NH–. They are distinguished from proteins on the basis of size, typically containing less than 50–100 monomer units. The shortest peptides are dipeptides, consisting of two amino acids joined by a single peptide bond.

The size boundaries which distinguish peptides, polypeptides, and proteins are arbitrary.

Long peptides such as amyloid beta can be considered proteins, whereas small proteins such as insulin can be considered peptides.

Peptides included in this overview have a minimum of two amino acids coupled together through an amide bond. No fixed maximum number of amino acids was defined, but for recombinant proteins a maximum length of 50 amino acids was imposed so that proteins such as insulin are not considered.

## 3 Market Overview

A market analysis delivered 51 products with a peptide as the active ingredient currently on the market.

Six products of therapeutic peptides reached global sales of over \$750 million each in 2008: Copaxone (glatiramer, an immunmodulator approved for the treatment of multiple sclerosis, \$3.2 billion), Lupron (leuprolide, gonadotropin-releasing hormone receptor agonist for the treatment of hormone-responsive cancers \$1.9 billion), Zoladex (goserelin, gonadotropin-releasing hormone receptor agonist for the treatment of hormone-responsive cancers, \$1.1 billion), Sandostatin (octreotide, an SST-2 and SST-5 receptor agonist approved for treatment of acromegaly, and symptoms associated with carcinoid syndrome and vasoactive intestinal peptide-secreting tumors \$1.1 billion), Forteo (\$780 million) and Byetta (\$750 million). Seven further peptide therapeutics have global annual sales between \$160 and \$750 million.

We found roughly 400 peptide therapeutics (about 75%), vaccines (20%), and diagnostics (5%) in the public domain most of which are still in the early clinical stages (I/II) and about 20 in clinical phase III. The most common indications are oncology and metabolic indications.

We found 28 peptide therapeutics approved for marketing in the US and other countries as well as 24 peptide therapeutics approved only outside the US.

The vast majority of peptides has a total length of 2–9 amino acids, while longer peptides seem to gain ground.

Some of the peptides were modified to alter their pharmacokinetic and pharmacodynamic profiles. The most common modifications found are pegylation; conjugation to albumin, lipids, antibodies or antibody fragments, and radiolabeling. The following table gives an overview of peptide drugs currently on the market.

Nearly all big pharmaceutical companies such as Novartis, Daiichi Sankyo, GSK, Sanofi-aventis, Roche, and Merck-Serono conduct phase III studies with product candidates based on active peptides.

# 4 Therapeutic Peptides in Patents

In a survey carried out in the DWPI database[1] we found 126 patent families on therapeutic peptides with priorities before 2000, starting around 1987. For priorities between 2000 and 2011, 1,063 DWPI families can be found, out of which 973 are PCT, US, or EP applications.

Surprisingly, for vaccines based on peptides 5,459 DWPI families exist as of the end of 2011, out of which 2,244 are based on MHC/HLA techniques. A total of 2,191 are for humans, out of which again 888 are focused on cancer and multiple sclerosis. Included in these numbers are applications which focus on adjuvants or other methods to enhance the properties (or delivery) of vaccines.

No trends could be observed with respect to the number of patent families covering therapeutic peptides or vaccines. Even for launched products (see above) there are cases with just one patent family protecting the product, but also other cases with more than 30 patent families associated to them.

# 5 Peptides in Vaccines

As seen in the sheer number of patents and applications for peptide vaccines, the application of peptides as vaccines seems to be the most promising approach. Nearly all applications and patents found were filed after the year 2000. This is mostly due to the fact that the mechanism behind the immunological reactions to peptides was only described in the 1990s by Prof. Rammensee from the University in Tübingen, Germany.[2]

---

[1] Derwent World Patents Index, which is a database containing patent applications and grants from 44 of the world's patent issuing authorities.

[2] Rammensee et al. (1993), (1997a, b), (1999).

20                                                                                        W. Flasche

**Table 1** Peptide drugs currently on the market

| Company | Key IP right covering the product[a] | Generic name | US brand name | Therapeutic category | Year of first US (EU[b]) approval |
|---|---|---|---|---|---|
| Teva | WO2005117902 | Glatiramer acetate | COPAXONE® | Allergy and Immunology | 1996 (2001 UK) |
| COR Therapeutics | WO09015620 | Eptifibatide | INTEGRILIN® | Cardiovascular | 1998 (1999) |
| Novo Nordisk | WO08202901 | Glucagon | GLUCAGEN® | Metabolic | 1998 (1992) |
| Senetek | WO09104039 | Aviptadil | INVICORP® | Urology | Last reported status phase III in US (1998 Denmark) |
| Roche | EP00277829 | Ganirelix acetate | ANTAGON® | Fertility | 1999 (2000) |
| Debio Recherche | US5192741 | Triptorelin | TRELSTAR DEPOT® | Oncology | 2000 (1995) |
| ASTA Medica | EP02266567 | Cetrorelix acetate | CETROTIDE® | Fertility | 2000 (1999) |
| Biogen Inc (licensed to Medicines Company) | WO09102750 | bivalirudin | ANGIOMAX® | Hematology | 2000 (2004) |
| Scios (J & J) | EP02098866 | nesiritide | NATRECOR®, NORATAK® | Cardiovascular | 2001 (discontinued in Europe) |
| Ferring | WO2005070449 | atosiban | TRACTOCILE® | Gynacology | Discontinued in US (2000 UK) |
| Lilly | US20040033950 | teriparatide | FORTEO® | Bones and Connective Tissue (osteoporesis) | 2002 (2003) |
| Hoffman-LaRoche | WO2007085567 | enfuvirtide | FUZEON® | Infection | 2003 (2003) |
| Praecis | WO2004056388 | abarelix | PLENAXIS® | Oncology | 2003(discontinued) (2005) |
| Elan | WO09107980 | ziconotide | PRIALT® | Pain | 2004 (2005) |
| Amylin | WO2010118384 | Pramlintide acetate | SYMLIN® | Metabolic | 2005 (No further development reported in the EU since 2001) |
| Amylin (cooperation with Lilly) | WO2011156407 | exenatide | BYETTA® | Metabolic | 2005 (2006) |

(continued)

**Table 1** (continued)

| Company | Key IP right covering the product[a] | Generic name | US brand name | Therapeutic category | Year of first US (EU[b]) approval |
|---|---|---|---|---|---|
| Beaufour Ipsen | WO2011117851 | lanreotide | SOMATULINE DEPOT® | Endocrinology | 2007 (1994 France) |
| Amgen | WO00024770 | romiplostim | NPLATE® | Hematology | 2008 (2009 EU) |
| Shire (through acquisition of Jerini) | EP00370453 | icatibant | FIRAZYR® | Cardiovascular | 2011 (2008) |
| Ferring | WO09846634 | degarelix | FIRMAGON® | Female infertility and Oncology | 2008 (2009) |
| Takeda (IDM) | WO09935162 | mifamurtide | JUNOVAN® | Oncology | Phase III in US (2009) |
| Novo Nordisk (license from Scios and Mass. Gen. Hosp.) | WO2007113205 | liraglutide | VICTOZA® | Metabolic | 2010 (2009) |

[a] As found in the FDA orange book and publications on the patent status by the companies
[b] Mentioned is the first country. If no country is mentioned the registration/launch was EU-wide

In short there are two classes of MHC-molecules: MHC class I molecules that can be found on most cells having a nucleus which presents peptides that result from proteolytic cleavage of mainly endogenous, cytosolic or nuclear proteins, DRIPS, and larger peptides. However, peptides derived from endosomal compartments or exogenous sources are also frequently found on MHC class I molecules. This non-classical way of class I presentation is referred to as cross-presentation in the literature. MHC class II molecules can be found predominantly on professional antigen presenting cells (APCs), and present predominantly peptides of exogenous proteins that are taken up by APCs during the course of endocytosis, and are subsequently processed. As for class I, alternative ways of antigen processing are described that allow peptides from endogenous sources to be presented by MHC class II molecules (e.g. autophagocytosis). Complexes of peptide and MHC class I molecule are recognized by CD8-positive cytotoxic T-lymphocytes bearing the appropriate TCR, complexes of peptide and MHC class II molecule are recognized by CD4-positive helper T-cells bearing the appropriate TCR.

CD4-positive helper T-cells play an important role in orchestrating the effector functions of antitumor T-cell responses and for this reason the identification of CD4-positive T-cell epitopes derived from Tumor-Associated Antigens (TAA) seem to be of great importance for the development of pharmaceutical products for triggering anti-tumor immune responses.[3]

It was shown in mammalian animal models, e.g., mice, that even in the absence of CTL effector cells (i.e., CD8-positive T lymphocytes), CD4-positive T-cells are sufficient for inhibiting manifestation of tumors via inhibition of angiogenesis by secretion of interferon-gamma (IFN$\gamma$).[4] Additionally, it was shown that CD4-positive T-cells recognizing peptides from tumor-associated antigens presented by HLA class II molecules can counteract tumor progression via the induction of an antibody (Ab) responses.[5]

In the absence of inflammation, expression of MHC class II molecules is mainly restricted to cells of the immune system, especially APCs, e.g., monocytes, monocyte-derived cells, macrophages, dendritic cells. In tumor patients, cells of the tumor have surprisingly been found to express MHC class II molecules.[6]

For a peptide to trigger (elicit) a cellular immune response, it must bind to an MHC-molecule. This process is dependent on the allele of the MHC-molecule and specific polymorphisms of the amino acid sequence of the peptide. MHC-class-I-binding peptides are usually 8–10 amino acid residues in length and usually contain two conserved residues ("anchor") in their sequence that interacts with the corresponding binding groove of the MHC-molecule. In this way each MHC allele has a "binding motif" determining which peptides can bind specifically to the binding groove.

---

[3] Kobayashi et al. (2002); Gnjatic et al. (2003); Qin et al. (2003).
[4] Qin and Blankenstein (2000).
[5] Kennedy et al. (2003).
[6] Dengjel et al. (2006).

In MHC-dependent immune reaction, peptides not only have to be able to bind to certain MHC molecules expressed by tumor cells, they also have to be recognized by T-cells bearing specific T-cell receptors (TCR).

The antigens that are recognized by the tumor-specific T-lymphocytes, that is, their epitopes, can be molecules derived from all protein classes, such as enzymes, receptors, transcription factors, etc. Furthermore, tumor-associated antigens, for example, can also be present in tumor cells only, for example as products of mutated genes. Another important class of tumor-associated antigens are tissue-specific antigens, such as CT ("cancer testis")-antigens that are expressed in different kinds of tumors and in healthy tissue of the testis.

Various tumor-associated antigens have been identified. Further, much research effort is expended to identify additional tumor-associated antigens. Some groups of tumor-associated antigens, also referred to in the art as tumor-specific antigens, are tissue specific. Examples include, but are not limited to, tyrosinase for melanoma, PSA and PSMA for prostate cancer and chromosomal cross-overs (translocations) such as bcr/abl in lymphoma. However, many tumor-associated antigens identified occur in multiple tumor types, and some, such as oncogenic proteins and/or tumor suppressor genes which actually cause the transformation event, occur in nearly all tumor types. For example, normal cellular proteins that control cell growth and differentiation, such as p53 (which is an example for a tumor suppressor gene), ras, c-met, myc, pRB, VHL, and HER-2/neu, can accumulate mutations resulting in upregulation of expression of these gene products thereby making them. These mutant proteins can also be a target of a tumor-specific immune response in multiple types of cancer.

Immunotherapy in cancer patients aims at activating cells of the immune system specifically, especially the so-called cytotoxic T-cells (CTL, also known as "killer cells", also known as CD8-positive T-cells), against tumor cells but not against healthy tissue. Tumor cells differ from healthy cells by the expression of tumor-associated proteins. HLA molecules on the cell surface present the cellular content to the outside, thus enabling a cytotoxic T cell to differentiate between a healthy and a tumor cell. This is realized by breaking down all proteins inside the cell into short peptides, which are then attached to HLA molecules and presented on the cell surface. Peptides that are presented on tumor cells, but not or to a far lesser extent on healthy cells of the body, are called tumor-associated peptides (TUMAPs).

For proteins to be recognized by cytotoxic T-lymphocytes as tumor-specific or -associated antigens, and to be used in a therapy, particular prerequisites must be fulfilled. The antigen should be expressed mainly by tumor cells and not by normal healthy tissues or in comparably small amounts. It is furthermore desirable, that the respective antigen is not only present in a type of tumor, but also in high concentrations (i.e. copy numbers of the respective peptide per cell). Tumor-specific and tumor-associated antigens are often derived from proteins directly involved in transformation of a normal cell to a tumor cell due to a function, e.g., in cell cycle control or apoptosis. Additionally, also downstream targets of the proteins directly causative for a transformation may be upregulated und thus be indirectly tumor associated. Such indirectly tumor-associated antigens may also be targets of a vaccination approach. Essential is in both cases the presence of epitopes in the amino acid sequence of the antigen, since such peptide ("immunogenic peptide") that is derived from a tumor-associated antigen should lead to an in vitro or in vivo T-cell response.

Basically, any peptide able to bind an MHC molecule may function as a T-cell epitope. A prerequisition for the induction of an in vitro or in vivo T-cell-response is the presence of a T-cell with a corresponding TCR and the absence of tolerance for this particular epitope. T-helper cells play an important role in orchestrating the effector function of CTLs in anti-tumor immunity. T-helper cell epitopes that trigger a T-helper cell response of the TH1 type support effector functions of CD8-positive killer T-cells, which include cytotoxic functions directed against tumor cells displaying tumor-associated peptide/MHC complexes on their cell surfaces. In this way tumor-associated T-helper cell peptide epitopes, alone or in combination with other tumor-associated peptides, can serve as active pharmaceutical ingredients of vaccine compositions which stimulate anti-tumor immune responses.

Since both types of responses, CD8 and CD4 dependent, contribute jointly and synergistically to the anti-tumor effect, the identification and characterization of tumor-associated antigens recognized by either CD8 + CTLs (MHC class I molecule) or by CD4-positive CTLs (MHC class II molecule) are important in the development of tumor vaccines. It is therefore an object of the present invention, to provide compositions of peptides that contain peptides binding to MHC complexes of either class.

The first clinical trials using tumor-associated peptides were started in the mid-1990s by Boon et al. mainly for the indication melanoma. Clinical responses in the best trials ranged from 10 to 30%. Severe side effects or severe autoimmunity have not been reported in any clinical trial using peptide-based vaccine monotherapy.

However, priming of one kind of CTL is usually insufficient to eliminate all tumor cells. Tumors are very mutagenic and thus able to respond rapidly to CTL attacks by changing their protein pattern to evade recognition by CTLs. To counterattack the tumor evasion mechanisms a variety of specific peptides are used for vaccination. In this way a broad simultaneous attack can be mounted against the tumor by several CTL clones simultaneously. This may decrease the chances of the tumor to evade the immune response. This hypothesis has been recently confirmed in a clinical study against renal cell carcinoma.

Patients suffering from renal cell carcinoma were treated with a vaccine composed of 13 different peptides.[7]

The major task in the development of a tumor vaccine is therefore not only the identification and characterization of novel tumor-associated antigens and immunogenic T-helper epitopes derived thereof, but also the combination of different epitopes to increase the likelihood of a response to more than one epitope for each patient.

---

[7] Singh-Jasuja et al. (2007); Staehler et al. (2007); Singh et al. (2010); Reinhardt et al. (2010).

SEQ ID NO. 5

Qy          1 GLWHHQTEV 9

              |||||||||

Db          676 GLWHHQTEV 684

**Fig. 1** Alignment taken from an office action received from the USPTO

# 6 Alignment

To determine whether peptides are similar to each other or a peptide disclosed in the art falls within the range of a disclosed and claimed homolog, the peptides have to be aligned.

This is done by a comparison of the two sequences. So far so good. Unfortunately, we find that many times the alignments done by the PTOs are simply wrong.

I will demonstrate this with a short example:

A hypothetical claim reads as simple as "A peptide consisting of an amino acid sequence according to SEQ ID No. 5" (see Fig. 1).

The short comment by the respective examiner was: "... amino acids 676–684 of SEQ ID NO. 9084 read on applicant's SEQ ID NO. 5".

There are some things that instantly come to mind:

- Consisting of means that "amino acids 676–687 of SEQ ID NO. 9084" is at least 675 amino acids too long, to fall under the claim.
- "SEQ ID NO. 9084", so finding anything in at least 9083 sequences is nearly impossible.
- Most important is that the alignment is wrong. If you have the consisting of language in the claim, the alignment must be made with the first amino acids in any given sequence. So in this example the alignment would have been correctly done with the amino acids 1–9 of SEQ ID NO. 9084. Even if the wording of the claim is more open, for example "consisting of SEQ ID NO. 5 and up to ten amino acids added to the C-terminus and/or the N-terminus of SEQ ID NO. 5" one cannot start the alignment anywhere in any given peptide or protein. The alignment always must start with the position 1 of any given peptide/protein.
- There is only one wording, that allows for an alignment as done by the PTO in the example: "a peptide or protein comprising a continuous stretch of amino acids according to (or given in) SEQ ID NO. 5[8]". In this example even a claim wording like "a peptide of up to 100 (200) amino acids comprising a continuous stretch of amino acids according to (or given in) SEQ ID NO. 5" would not allow for the alignment as done by the PTO.

---

[8] I believe all readers know that there are variants of that claim language which can be applied, but those variants are a mere linguistic variations.

# 7 The Skilled Artisan

Many decisions in the major jurisdictions exist which relate to the question of who "one skilled in the art" really is. One particular issue becomes obvious when dealing with therapeutic peptides, and maybe also in other areas in the biotech field.

Even topics that are discussed in textbooks for undergraduates are often said to be a burden for one skilled in the art, or unclear, or not properly disclosed, the reasons go on and on. In many cases we had to hand in extracts from textbooks— for example Janeway's Immunobiology[9]—to the PTO, just to explain the basic techniques to the examiner. Just to pinpoint the problems associated with rejections based for example on "unpredictability" or the understanding of the examiner. "Janeway's Immunobiology" is a textbook that introduces the immune system in all its aspects to undergraduates. Unfortunately, we are far from the discussion led in other areas on who one skilled in the art is.

It seems that particularly the United States Patent and Trademark Office (USPTO) issues office actions based on so-called "112 rejections",[10] i.e., based on the allegation that the patent specification does not contain a sufficient written description of the invention as to enable any person skilled in the art to make and use the invention. One can only hope that the situation will get better in the future.

# 8 Fusion Proteins

In some cases and for some applications it is useful to fuse peptides to proteins, to parts of proteins, or a long polypeptide. For example the proteins can act as a carrier or stabilize the peptide in vivo.

There is one aspect to keep in mind. For example the disclosure of WO2010117760 reads as:

> In one aspect, the present invention is directed to fusion proteins that comprise one or more therapeutic peptides bound to an Fe domain by a glycine succinate linker. As contemplated herein, when a glycine succinate linker is used to link a therapeutic peptide and a canine Fe domain, the glycine residue of the linker is linked to the N- terminus of the Fe domain and the succinate moeity is linked to the C-terminus of the therapeutic peptide, and/or an amino acid linker of various length and sequence. In relation to the linker, the length and composition are necessary to achieve prolonged efficacy of the therapeutic peptide. As contemplated herein, the therapeutic peptide may be linked to the Fe domain in different orientations. In one orientation, the C-terminus of peptide is linked to the N-terminus of the Fe domain and in another orientation, the N-terminus of the peptide is linked to the N-terminus of the Fe domain. The Fe domain exists as a homodimer of the

---

[9] Janeway's Immunobiology, 8th edition (1. August 2011), Taylor and Francis ISBN-13: 978-0815342434.

[10] The term alludes to 35 U.S.C. 112, which defines the requirements for a patent specification.

hinge, $CH_2$ and $CH_3$ regions of an IgG molecule, with the Fe domain beginning at the first N-terminal cysteine residue within the IgG hinge region and the homodimer is held together by two disulfide bonds in the hinge from the cysteine residues therein.

Very often, patent authorities issue rejections like the one shown because either the sequence of the protein part or the way the protein and the peptide are fused are allegedly not sufficiently disclosed.

This can happen even if the protein is properly deposited. In any case the disclosure should always include different ways of describing how the parts are fused, what length the fusion protein has, and all other possible information. Although this has been considered in the cited example, some examiners might still question which amino acid of the peptide is fused to which amino acid of the protein, or how long the protein is, what continuous stretch of amino acids is meant, and the like.[11] Specific examples follow in the application used as an example. If the quoted paragraph was the only disclosure of the fusion, it would be problematic. In WO2010117760 the sequence of the Fe domain is explicitly disclosed but with more than one possible sequence. Besides the problematic wording of "the Fe domain" for more than one possible sequence this plurality directly leads to the problem of unity.

# 9 Unity of the Invention

The unity requirement is set forth in Rule 13 of the Patent Cooperation Treaty[12] (the same wording can be found in Art. 82 EPC). According to Rule 13.1 a patent application shall relate to one invention only or to a group of inventions so linked to form a single general inventive concept. According to 13.2 the requirement of unity of invention shall be fulfilled in an application covering more than one invention only when there is a technical relationship among those inventions involving one or more of the same or corresponding special technical features. The expression "special technical features" shall mean those technical features that define a contribution which each of the claimed inventions, considered as a whole, makes over the prior art.

We discuss this issue here because the patent offices apply different requirements when considering unity of invention in peptide applications than in small molecule applications.

---

[11] In the disclosure of WO2010117760 more specific embodiments are described. The shown paragraph was only chosen to demonstrate some general problems.

[12] The Patent Cooperation Treaty (PCT) is a treaty which provides a unified procedure for filing and examining patent applications to protect inventions in each of its contracting states. A set of Rules exist which sets forth basic standards for patentability, unity of a patent application, and the like. See http://www.wipo.int/pct/en/texts/rules/rtoc1.htm.

## 9.1 Markush Claims

In an interesting article Leber[13] discussed the use of Markush claims in biotechnology under PCT and EPC guidelines. In short, Markush claims to cover different embodiments (solutions) of an invention that have a common "core".

> The requirement of unity is reflected in both the Patent Cooperation Treaty (PCT) and European Patent Convention (EPC). The relevant articles, rules and guidelines (GL) are highly similar in these two bodies of law, resulting in a uniform assessment of unity before the EPO and international authorities under the PCT (Case Law of the Boards of Appeal of the European Patent Office, 5th ed. 2006, II.C.1).

Leber concludes that

> claims referring to a plurality of alternative nucleic acids or proteins can be worded as Markush formulae and his detailed analysis of Rule 13.2 PCT (Rule 44(1) EPC) resulted in the conclusion that unity of a Markush claim requires that a significant common structural element, which may be the common core structure of the Markush formula or a part thereof, must contribute either as such or in combination with the respective variable parts of the Markush formula to a special technical feature that represents a contribution over the prior art, i.e., is novel and inventive.
>
> Neither the examples on chemical-type Markush claims in the PCT guidelines nor the general introduction into unity of Markush claims in the PCT and EPC guidelines, nor the related part in the PCT Administrative Instructions are in line with this requirement. In view of the correspondence between the articles, rules, and guidelines of the PCT and EPC on unity matters, the present analysis suggests that the guidelines of the EPC on the unity of Markush claims are also not in line with the articles and rules of the EPC.

In other words: The possibility to use Markush claims differ enormously between "chemical claims" and "BioTech" claims. This is also what we see in the communications under PCT or from the European Patent Office (EPO).

So let us take a step back and have a look at the differences between DNA, peptides and "chemical", and the implications for unity.

DNA (or RNA and other synthetic variants) is commonly depicted by four letters A, T, C, and G which represent the four nucleotides Adenine, Thymine, Cytosine, and Guanine which are connected by covalent bonds starting at the $3'$ end of the sequence.

An example would be: $3'$-GGG AAA TTT CCC GAG AGA TCT-$5'$. Omitted in this formula are the sugar and phosphate residues that build the backbone of the fragment. DNA is only a linear polymer of covalently linked nucleotide monomers.

Therefore, two "cores" could be defined for a Markush group. The first core for example could be "AAA TTT CCC GAG AGA":

$$3' - GGG\ \textbf{AAA TTT CCC GAG AGA}\ TCT - 5'$$
$$3' - GAG\ \textbf{AAA TTT CCC GAG AGA}\ TAT - 5'$$
$$3' - GTG\ \textbf{AAA TTT CCC GAG AGA}\ TTT - 5'$$

---

[13] Leber (2009).

**Fig. 2**  Hypothetical heterocyclic compound

A Markush-type claim could be: The nucleic acid as specified by $3'$-$R_1$-**AAA TTT CCC GAG AGA**-$R_2$-$5'$ with $R_1 = G\underline{A}G$ and $R_2 = T\underline{A}T$ or $R_1 = G\underline{T}G$ and $R_2 = T\underline{T}T$.

Such a claim would be clear to a person skilled in the art and would have the same scope as a claim reciting the three alternatives.

The second "core" could be the sugar and phosphate residues of the nucleotides (N) with the actual base being the exchangeable part. $3'$-$N(R_1)$ $N(R_1)$ $N(R_1)$ $N(R_2)$ $N(R_2)$ $N(R_2)$ $N(R_3)$ $N(R_3)$ $N(R_3)$ $N(R_4)$ $N(R_4)$ $N(R_4)$ $N(R_1)$ $N(R_2)$ $N(R_1)$ $N(R_2)$ $N(R_1)$ $N(R_2)$ $N(R_3)$ $N(R_2)$ $N(R_3)$-$5'$ with $R_1 = g$ $R_2 = a$ $R_3 = t$ and $R_4 = c$. As it can be easily seen this more chemical view (omitting the real structure and not defining the middle part as done above) is not practicable. But what can be seen is that the compound really is a polymer with narrowly defined residues.

Nevertheless, the PCT guidelines[14] dismiss this. In paragraph 10.52, which relates to unity of invention in Biotechnological Inventions, the following can be found:

> The sugar-phosphate backbone cannot be considered a significant structural element, since it is shared by all nucleic acid molecules.

And there is another layer. When looking at a sequence of DNA (or RNA) the examiner and the person skilled in the art is looking at the information incorporated in the disclosed sequence. Three nucleotides code for an amino acid and some amino acids are coded by more than one combination. So some changes in the sequence do not lead to any change in the coded amino acid, while other changes may lead to big changes as for example a small non-polar amino acid would be changed into a big polar amino acid.

How does this compare to "chemical" compounds ? A typical structure for a pharmaceutical active compound claimed is shown in Fig. 2 (fictional, not drawn to an existing application): wherein $X = NR_3$, O or S with $R_1 =$ aryl, alkyl or alkene $R_2 =$ alkyl and $R_3 =$ aryl.

The core would be the two ring system which is known for a long time. The defined substitutions of X and $R_1$ to $R_3$ lead to billions of possible combinations and resulting compounds all falling under the scope of the claim. Even if the residues R1–R3 would be narrower defined—let us say instead of "alkyl" as it

---

[14] PCT International Search and Preliminary Examination Guidelines, which are non-binding guidelines for searching and examining PCT applications applied by examiners for the different search and examination authorities. Available online under http://www.wipo.int/pct/en/texts/pdf/ispe.pdf.

stands, alkyl with 6 or less C-Atoms, or even just methyl to butyl—the number of possible compounds is still overwhelmingly large.

The ring system will be considered as a significant structural element, whether or not ring systems as in the example are known or shared by organisms in one way or the other (see also example 18 of the PCT guidelines).

Obviously, there is no "information" in the compounds as is the case with DNA/RNA.

## 9.2 What About Peptides?

Peptides are also linear[15] polymers of covalently linked amino acid monomers. There are 20 so-called proteinogenic amino acids that can be found in proteins and require cellular machinery coded for in the genetic code of any organism for their isolated production. There are 22 standard amino acids, but only 21 are found in eukaryotes. Of the 22, 20 are directly encoded by the universal genetic code.[16]

They can be divided into several groups based on their properties. Important factors are charge, hydrophilicity or hydrophobicity, size, and functional groups. Amino acids are commonly depicted by a one-letter (or three-letter) system. For example the non-polar amino acid Alanine is depicted A (Ala).

So let us take for example the 10mer HIL**ARND**HIL, to which the following derivatives exist:

KMF**ARND**KMF

TWV**ARND**TWV

LKF**ARND**LKF

A Markush claim could be: A peptide with 10 consecutive amino acids specified by R-ARND-R with R selected from the group consisting of HIL, KMF, TWV, or LKF.

Such a claim would be clear to a person skilled in the art and would have the same scope as a claim reciting the three alternatives.

Although the differences between the four peptides seem to be bigger as in the example of DNA HIL, KMF, and TWV all show a sequence of polar–nonpolar–nonpolar, whereas LKF is nonpolar–polar–nonpolar. So, besides a possibility to define the peptides also through the properties of the amino acids, the differences are not as big as in the example for a "chemical" compound. There are just 20 residues that can be interchanged and as they are more likely to be changed to amino acids with comparable properties the possibilities are even more reduced. At least for peptides with a length of up to 50 amino acids the possible number of

---

[15] There are circular peptides, but they are disregarded at this point of the discussion.

[16] Ambrogelly et al. (2007).

alternatives (with a core) is much more limited than with the example of a traditional small molecule.

The PCT guidelines give some examples[17] on the unity of biotechnological inventions such as peptides or nucleotides. Examples 18–24 are drawn from "chemical" Markush groups and Examples 32–37 from "BioTech" Markush groups.

Chemical Markush groups need a "significant structural element that is shared by all alternatives" and all alternatives must possess a common property.

The situation is completely different for "BioTech" Markush groups:

- The examples given propose an assessment of whether or not the significant common structural element is essential to the common property. See examples 33–37. In example 33 "the description discloses that SEQ ID NOs.1–10 share a common property, that is, expression of an mRNA present only in patients afflicted with disease Y. Moreover, SEQ ID NOs. 1–10 share a significant structural element that is essential to the common property, i.e., a probe comprising the shared structural element can detect the mRNA of patients afflicted with disease Y. Since both of these requirements are met, the group of polynucleotide molecules claimed meets the requirement of unity of invention (a priori)."

- In example 33 unity is assessed a priori. In examples 34–37 unity is assessed a posteriori. There is no equivalent assessment in chemical cases.

Leber discusses this using example 35:

BiotechType example 35 describes, in brief, the following problem: A claim refers to a fusion protein comprising carrier protein X linked to a polypeptide selected from SEQ ID Nos. 1, 2, or 3. The description discloses that carrier protein X is 1000 amino-acids in length and functions to increase the stability of the fusion proteins in the blood stream. SEQ ID Nos. 1, 2, and 3 are small peptide epitopes (10–20 residues in length) isolated from different antigenic regions of E.coli. SEQ ID Nos. 1, 2, and 3 do not share any significant common structure. Both the structure of protein X and its function as a carrier protein are known in the prior art.

The alternative compounds claimed in this example thus share a significant common structural element, namely protein X. Moreover, the compounds as a whole all have the same property in that they serve as a vaccine to E.coli. This property is only related to the antigenic regions of E.coli and not to the significant common structural element protein X.

Following the instructions derivable from the ChemType Markush examples, the above claim would be unitary as there is a significant common structural element, i.e., protein X, and as the compounds as a whole have a common property/activity, i.e., being a vaccine to E. coli.

According to BiotechType example 35, however, the above claim lacks unity because the significant common structural element, i.e., protein X, neither on its own nor in combination with any of SEQ ID Nos. 1, 2, or 3 represents a special technical feature (Rule 13.2 PCT) as the common property is not linked to the significant common structural element protein X but only to the peptides of SEQ ID Nos. 1, 2, and 3, and further, because protein X and its function are known in the art.

---

[17] PCT guidelines, paragraphs 10.38–10.57.

This is not a theoretical problem. Here is an example from the specification of WO2006031727 titled "Peptides for Treatment of Autoimmune Diseases":

An embodiment of the present invention is a composition having a peptide comprising an amino acid sequence selected from the group consisting of: YEAYK (SEQ ID NO: 19), FEAYK (SEQ ID NO: 20), EEAYK (SEQ JX) NO: 21), VEAYK (SEQ ED NO: 22), EEAFK (SEQ ID NO: 23), FEAFK (SEQ ID NO: 24), VEAFK (SEQ ID NO: 25), and YEAFK (SEQ ED NO: 26). The peptide comprising any of these sequences is in some embodiments at least six amino acids in length, or at least 15 amino acids in length.

These sequences are motifs that will be combined with other motifs. The possibility of a Markush claim would help to keep the disclosure short and clear as well as the costs for the applicant low. The application goes on like this:

Another embodiment is a composition having a peptide comprising an amino acid sequence selected from the group consisting of: EKAKYEAYKAAAAAA (SEQ ID NO: 1), EKPKYEAYKAAAAPA (SEQ ID NO: 2), EKPKFEAYKAAAAPA (SEQ ED NO: 3), EKAKEEAYKAAAAAA (SEQ ED NO: 4), EKPKVEAYKAAAAPA (SEQ ID NO: 5), EKPKEEAFKAAAAPA (SEQ ID NO: 6), EKAKFEAFKAAAAAA (SEQ TD NO: 7), APEKAKFEAFKAAAAPA (SEQ ED NO: 8), APEKAKFEAYKAAAAPA (SEQ ED NO: 9), APEKAKVEAFKAAAAPA (SEQ ID NO: 10), EAKKYEAYKAAAAAA (SEQ ID NO: 11), EAPKFEAYKAAAAPA (SEQ ED NO: 12), EAPKVEAYKAAAAPA (SEQ ID NO: 13), EAPKFEAFKAAAAPA (SEQ ED NO: 14), APEAKKFEAFKAAA-APA (SEQ ID NO: 15), APEAKKFEAYKAAAAPA (SEQ ID NO: 16), and APEAKKVE AFKAAAAPA (SEQ ED NO: 17). Yet another embodiment of the present invention is a composition having a peptide comprising an amino acid sequence selected from the group consisting of: EKAKYEAYK (SEQ ID NO: 27), EKPKYEAYK (SEQ ED NO: 28), EKPKFEAYK (SEQ ID NO: 29), EKAKEEAYK (SEQ ID NO: 30), EKPKVEAYK (SEQ ID NO: 31), EKPKEEAFK (SEQ DD NO: 32), EKAKFEAFK (SEQ DD NO: 33), EKAKFEAYK (SEQ ID NO: 34), EKAKVEAFK (SEQ ED NO: 35), EAKKYEAYK (SEQ ED NO: 36), EAPKFEAYK (SEQ DD NO: 37), EAPKVEAYK (SEQ ED NO: 38), EAPKFEAFK (SEQ ID NO: 39), EAKKFEAFK (SEQ DD NO: 40), EAKKFEAYK (SEQ DD NO: 41), and EAKKVEAFK (SEQ DD NO: 17).

While there is no unity problem before the EPO or under PCT (and in many other jurisdictions) in case, e.g., a cell expressing DNA (or for producing a peptide), the DNA coding for a peptide and the peptide as well is claimed, unity is an important issue in the US.

Office actions in which 100 or more groups of invention are identified and a restriction is required are not unheard of (see the chapter on the situation in the US).

The underlying and most important question is: Is there any justification for applying the rules differently in different technical fields?

The reader might ponder why this inconsistency has to be in the field of peptides/biotech, but unfortunately for all professionals in patents other inconsistencies can be found in other fields.

EPO Technical Board decision T 777/08,[18] which is the subject of quite some discussions,[19] states that a claim to a crystalline form or polymorph of a compound

---

[18] http://www.epo.org/law-practice/case-law-appeals/recent/t080777ex1.html

[19] http://ipkitten.blogspot.com/2012/02/what-is-obvious-route-or-destination.html

Atorvastatin), which is defined, for example, by its X-ray diffraction data, will in general lack inventive step over prior art disclosing the amorphous form of the same compound, unless in any particular case there is a technical prejudice or unexpected property of the polymorph discovered. The main reason given was that screening for polymorphs is a routine procedure.

Note that—again—the technical board refers to the—in this decision unexpected—properties; this time within the field of small molecules.

The counter view, following the reasoning to which some practitioners have been accustomed, is that although it may be routine to look for crystalline forms, there is absolutely no way to predict what crystalline form(s) will be found, and therefore any specifically defined crystalline form is not obvious.

In the case of polymorphs, it might now be a routine procedure for the skilled person to conduct a polymorph screen using the available techniques. The skilled person "would do" a polymorph screen and "would reasonably expect" to find new polymorphs. These polymorphs would be expected to have different properties in relation to solubility, stability, etc. The skilled person is looking for such properties to improve and therefore will look for the polymorph having the best profile overall. But is it really a general lack inventive step using standard techniques to find a solution to a problem, as long as a person skilled in the art cannot predict the outcome of such experiments? How is the invention defined? If it is including the structural features (as in small molecules), is it obvious? This is up for discussion.

Another technical field with a similar inconsistency is antibodies (see corresponding chapter in this volume). After Technical Board decision T735/00[20] the EPO only acknowledges inventive step for an antibody (claimed structurally) "if and when there is evidence that a claimed monoclonal antibody prepared by routine methods shows unexpected properties". This is inconsistent with the approach taken in the field of small molecule chemistry.[21]

# 10 Definitions and Problems Related to Definitions

According to a German proverb, a patent is its own encyclopedia. This means that definitions play a paramount role to make clear what is meant with the used claim langiage, both during prosecution (where clarity is needed to determine whether the claimed subject matter is novel) and enforcement (where clarity is needed to determine whether a competing embodiment infringes the patent. In biosequence claims, some specific terms are frequently used which will be explained in the following.

---

[20] http://www.epo.org/law-practice/case-law-appeals/recent/t000735eu1.html
[21] Stewart et al. (2011).

## 10.1 Homology

An example for a broader definition of peptides to not restrict the protection solely
on the peptide itself is:

> The present invention further relates to a peptide comprising at least one sequence selected
> from the group consisting of SEQ ID No. 1 to SEQ ID No. 16, or a variant thereof which is
> at least 85% homologous to SEQ ID No. 1 to SEQ ID No. 5 and induces mammalian
> T cells cross-reacting with said sequences, wherein said peptide is not the full-length
> peptide of SEQ ID No. 6.

In this way peptides comprising the recited sequences can be claimed as long as
they are not the full-length peptide (protein). Furthermore, also peptides with a
homology of at least 85% which show the same activity are claimed.

First of all the term "homologous" must be defined:

> In the present invention, the term "homologous" refers to the degree of identity between
> sequences of two amino acid sequences, i.e. peptide or polypeptide sequences. The
> aforementioned "homology" is determined by comparing two sequences aligned under
> optimal conditions over the sequences to be compared. The sequences to be compared
> herein may have an addition or deletion (for example, gap and the like) in the optimum
> alignment of the two sequences. Such a sequence homology can be calculated by creating
> an alignment using, for example analysis tools provided by public databases.

But while writing an application one should be aware what changes an 85%
homology allows. With DNA or proteins 85% homology allows dozens or hundreds of changes. If the claimed peptides have a length of 6–8 amino acids—as
most of the therapeutic peptides on the market or in clinical trials (see supra) 85%
homology allows only 1 changed peptide. In a 50mer up to 8 changes are possible.
So if dealing with short peptides the homology should either not be used or further
definitions should be included in the description.

## 10.2 Variants

Another popular way of defining possible changes or derivates of peptides is the
term "variant". An example corresponding to the above definition of homology
could read as follows[22]:

> By a "variant" of the given amino acid sequence the inventors mean that the side chains
> of, for example, one or two of the amino acid residues are altered (for example by
> replacing them with the side chain of another naturally occurring amino acid residue or
> some other side chain) such that the peptide is still able to bind to *the target* in sub-
> stantially the same way as a peptide consisting of the given amino acid sequence in SEQ
> ID NO: 1–16.

---

[22] Example used in patent applications drafted by the author.

For certain positions in the sequence these should also be defined accordingly. For example, a peptide may be modified so that it at least maintains, if not improves, the ability to induce mammalian T cells cross-reacting[23]:

> As can be derived from the scientific literature and databases (or the description itself) certain positions of the peptides are typically anchor residues forming a core sequence, which is defined by polar, electro-physical, hydrophobic and spatial properties of the polypeptide chains constituting the binding groove. Thus one skilled in the art would be able to modify the amino acid sequences set forth in SEQ ID NO:1 to SEQ ID NO:16, by maintaining the known anchor residues, and would be able to determine whether such variants maintain the ability to induce mammalian T cells cross-reacting with said sequences.

Still the description should include more information about the nature of the possible modifications. In many jurisdictions this is needed for the support of the claims and to be not rejected as unclear[24]:

> Those amino acid residues that do not substantially contribute to interactions with the T-cell receptor can be modified by replacement with another amino acid whose incorporation does not substantially affect T-cell reactivity and does not eliminate binding to the relevant receptor.
>
> Therefore the original peptides disclosed herein can be modified by the substitution of one or more residues at different, possibly selective, sites within the peptide chain, if not otherwise stated. Such substitutions may be of a conservative nature, for example, where one amino acid is replaced by an amino acid of similar structure and characteristics, such as where a hydrophobic amino acid is replaced by another hydrophobic amino acid. Even more conservative would be replacement of amino acids of the same or similar size and chemical nature, such as where leucine is replaced by isoleucine. In studies of sequence variations in families of naturally occurring homologous proteins, certain amino acid substitutions are more often tolerated than others, and these are often shown in correlation with similarities in size, charge, polarity, and hydrophobicity between the original amino acid and its replacement, and such is the basis for defining "conservative substitutions".
>
> Conservative substitutions are herein defined as exchanges within one of the following five groups: Group 1-small aliphatic, nonpolar or slightly polar residues (Ala, Ser, Thr, Pro, Gly); Group 2-polar, negatively charged residues and their amides (Asp, Asn, Glu, Gln); Group 3 -polar, positively charged residues (His, Arg, Lys); Group 4-large, aliphatic, nonpolar residues (Met, Leu, Ile, Val, Cys); and Group 5-large, aromatic residues (Phe, Tyr, Trp).
>
> Less conservative substitutions might involve the replacement of one amino acid by another that has similar characteristics but is somewhat different in size, such as replacement of an alanine by an isoleucine residue. Highly non-conservative replacements might involve substituting an acidic amino acid for one that is polar, or even for one that is basic in character. Such "radical" substitutions cannot, however, be dismissed as potentially ineffective since chemical effects are not totally predictable and radical substitutions might well give rise to serendipitous effects not otherwise predictable from simple chemical principles.
>
> Of course, such substitutions may involve structures other than the common L-amino acids. Thus, D-amino acids might be substituted for the L-amino acids commonly found in the peptides of the invention and yet still be encompassed by the disclosure herein. In addition, amino acids possessing non-standard R groups (i.e., R groups other than those

---

[23] Id.

[24] Id.

**Table 2** Fictional example of how to disclose variants of a given peptide sequence

| Name | Peptide code | | S | I | G | Q | N | I | Q | Q | V |
|---|---|---|---|---|---|---|---|---|---|---|---|
| SEQ ID X | Variants | X1 | | M | | | | | | | L |
| | | X2 | L | | | | | | | | L |
| | | X3 | | | | | | | | | K | |
| | | X4 | I | | A | G | I | | A | | |
| | | X5 | L | | Y | P | K | L | Y | | |
| | | X6 | F | | F | T | Y | T | H | | |
| | | X7 | | | P | | R | | | | |
| | | X8 | M | | | | F | | | | |
| | | Position | 1 | 2 | 3 | 4 | 5 | 6 | 7 | 8 | 9 |

found in the common 20 amino acids of natural proteins) may also be used for substitution purposes to produce immunogens and immunogenic polypeptides according to the present invention.

If substitutions at more than one position are found to result in a peptide with substantially equivalent or greater antigenic activity as defined below, then combinations of those substitutions will be tested to determine if the combined substitutions result in additive or synergistic effects on the antigenicity of the peptide.

Still, a limit of changes should be given. As for example[25]:

At most, no more than 4 positions within the peptide would simultaneously be substituted.

If there are tested variants that the applicant knows of, these should be disclosed explicitly in the description. This can be done for example in a table as in the fictional example shown in Table 2:

One should be aware that X4–X6 would not fit the description set out above ("At most, no more than 4 positions within the peptide would simultaneously be substituted."). Therefore a description is necessary stating how the table should be read. If the changes must always occur together or the situation is a "can" situation, so for example in X6 when in position 1 there is an "F" position, 3 can also be changed to "F".

One example for a good definition and a slightly different approach is WO03040309. The definitions are short enough but at the same time start at a very open reading and end with the important and therefore more concrete defined variants.

[39] Variants are Polypeptides of the invention that have or more amino acid sequence changes with respect to the amino acid sequences shown in SEQ ID NOS: 6–32. Variants also can have amino acids joined to each other by modified peptide bonds, i.e., peptide isosteres, and may contain amino acids other than the 20 naturally occurring amino acids. [40] Preferably, variants contain one or more conservative amino acid substitutions (i.e., 1, 2, 3, 4, 5, 6, 7, 8, 9, or 10 substitutions), preferably at nonessential amino acid residues. A "nonessential" amino acid residue is a residue that can be altered from a wild type sequence of a protein without altering its biological activity, whereas an "essential" amino

---

[25] Id.

acid residue is required for biological activity. A conservative amino acid Substitution is one in which the amino acid residue is replaced with an amino acid residue having a similar side chain. Families of amino acid residues having similar side chains have been defined in the art. These families include amino acids with basic side chains (e.g. lysine, arginine, histidine), acidic side chains (e.g. aspartic acid, glutamic acid), uncharged polar side chains (e.g. glycine, asparagine, glutamine, serine, threonine, tyrosine, cysteine), nonpolar side chains (e.g. alanine, valine, leucine, isoleucine, proline, Phenylalanine, methionine, tryptophan), beta-branched side chains (e.g. threonine, valine, isoleucine) and aromatic side chains (e.g. tyrosine, Phenylalanine, tryptophan, histidine). Non-conservative substitutions would not be made for conserved amino acid residues or for amino acid residues residing within a conserved protein domain.

[41] Conservative amino acid substitutions are preferably at positions 11, 12, 16, 17, or 18 of the consensus Polypeptide shown SEQ ID NO: 34. Position 11 preferably is R, S, A, K, G, or T and more preferably R, A, G, or S. Position 12 preferably is K, N, R, H, A, S or Q and more preferably K, A, S, or N. Position 16 preferably is K, R, V, I, L, M, F, W, Y, A, S, T, N, Q, G, or H, and more preferably K, V, I, F, A, S, or N. Position 17 preferably is D, E, H, K, R, F, I, L, M, Y, V, W, A, S, T, N, Q, or G, and more preferably R, A, L, M, V, S, H, E, or Q. Position 18 preferably is K, R, F, I, L, Y, V, M, A, G, or H and more preferably K, R, F, I, L, Y or A. All possible combinations of substitutions at positions 11, 12, 16, 17, and 18, including no Substitution at any one, two, three, or four of these positions, are specifically envisioned.

[42] Variants also include Polypeptides that differ in amino acid sequence due to mutagenesis. Variants that function as both GLP-1 receptor agonists and glucagon receptor antagonists can be identified by Screening combinatorial libraries of mutants, for example, mutants of Polypeptides with conservative substitutions at 1 or more positions (i.e., at 1, 2, 3, 4, 5, 6, 7, 8, 9, or 10 positions) can be screened for GLP-1 receptor agonist activity and glucagon receptor antagonist activity using methods well known in the art and described in the specific examples below.

## 10.3 Analogs

In WO03040309 there is also a very short definition of "Analogs".

[43] An analog includes a propolypeptide, wherein the propolypeptide includes an amino acid sequence of a Polypeptide of the invention. Active Polypeptides of the invention can be cleaved from the additional amino acids in the propolypeptide molecule by natural, in vivo processes or by procedures well known in the art, such as by enzymatic or chemical cleavage.

As long as the "propolypeptide" are not known and disclosed this definition is not very helpful. It might be worthwhile to include a similar definition, if, for example, short peptides are modified with amino acids at one or both terminals to enhance the solubility or stability in the body. If possible cleavage points should be disclosed or the general concept explained.

## 10.4 Chemical Modifications

As the "variants" more or less are only substitutions of the amino acids other modifications must also be defined. Even some of the marketed peptides show such modifications. The reasons the peptides were modified was mostly to alter their pharmacokinetic and pharmacodynamic profiles. The most common modifications found are pegylation; conjugation to albumin, lipids, antibodies or antibody fragments, and radiolabeling.

Each of the modification must be defined in the description. Failure to do so will most certainly lead to the rejection of the claims. As always, the definition should start with the general concept and define further possibilities in greater detail[26]:

> In addition, the present invention further provides a peptide according to the present invention as described herein, wherein said peptide comprises chemically modified amino acids, and/or includes non-peptide bonds.

or

> In addition, the peptide or variant may be modified further to improve stability and/or binding to the target molecules in order to show an effect. Methods for such an optimization of a peptide sequence are well known in the art and include, for example, the introduction of reverse peptide bonds or non-peptide bonds.

Now these bonds have to be defined[27]:

> In a reverse peptide bond amino acid residues are not joined by peptide (-CO-NH-) linkages but the peptide bond is reversed. Such retro-inverso peptidomimetics may be made using methods known in the art.[28] This approach involves making pseudopeptides containing changes involving the backbone, and not the orientation of side chains. Retro-inverse peptides, which contain NH-CO bonds instead of CO-NH peptide bonds, are much more resistant to proteolysis. A non-peptide bond is, for example, $-CH_2-NH$, $-CH_2S-$, $-CH_2CH_2-$, $-CH = CH-$, $-COCH_2-$, $-CH(OH)CH_2-$, and $-CH_2SO-$. United States Patent 4,897,445 provides a method for the solid phase synthesis of non-peptide bonds ($-CH_2-NH$) in polypeptide chains which involves polypeptides synthesized by standard procedures and the non-peptide bond synthesized by reacting an amino aldehyde and an amino acid in the presence of $NaCNBH_3$.

If the stability of the peptides is to be enhanced also these changes have to be defined:

> Peptides comprising the sequences described above may be synthesized with additional chemical groups present at their amino and/or carboxy termini, to enhance the stability, bioavailability, and/or affinity of the peptides. For example, hydrophobic groups such as carbobenzoxyl, dansyl, or t-butyloxycarbonyl groups may be added to the peptides' amino termini. Likewise, an acetyl group or a 9-fluorenylmethoxy-carbonyl group may be placed

---

[26] Example used in patent applications drafted by the author.

[27] Id.

[28] Meziere et al. (1997).

at the peptides' amino termini. Additionally, the hydrophobic group, t-butyloxycarbonyl, or an amido group may be added to the peptides' carboxy termini. Further, the peptides of the invention may be synthesized to alter their steric configuration. For example, the D-isomer of one or more of the amino acid residues of the peptide may be used, rather than the usual L-isomer. Still further, at least one of the amino acid residues of the peptides of the invention may be substituted by one of the well known non-naturally occurring amino acid residues. Alterations such as these may serve to increase the stability, bioavailability and/or binding action of the peptides of the invention.

Similarly, a peptide or variant of the invention may be modified chemically by reacting specific amino acids either before or after synthesis of the peptide.[29] Chemical modification of amino acids includes but is not limited to, modification by acylation, amidination, pyridoxylation of lysine, reductive alkylation, trinitrobenzylation of amino groups with 2,4,6-trinitrobenzene sulphonic acid (TNBS), amide modification of carboxyl groups and sulphydryl modification by performic acid oxidation of cysteine to cysteic acid, formation of mercurial derivatives, formation of mixed disulphides with other thiol compounds, reaction with maleimide, carboxymethylation with iodoacetic acid or iodoacetamide and carbamoylation with cyanate at alkaline pH, although without limitation thereto.[30]

If the applicant has certain modifications in mind or certain amino acids are more likely to be modified these should also be included in the description:

Briefly, modification of, e.g., arginyl residues in proteins is often based on the reaction of vicinal dicarbonyl compounds such as phenylglyoxal, 2,3-butanedione, and 1,2-cyclohexanedione to form an adduct. Another example is the reaction of methylglyoxal with arginine residues. Cysteine can be modified without concomitant modification of other nucleophilic sites such as lysine and histidine. As a result, a large number of reagents are available for the modification of cysteine.

Another example of a short and clear definition is:

"For example, diethylpyrocarbonate is a reagent for the modification of histidyl residues in proteins. Histidine can also be modified using 4-hydroxy-2-nonenal."

For the modification with PEG the following should be sufficient, although we found descriptions for more than a page even though the claimed invention did not use PEG modifications at all:

Successful modification of therapeutic proteins and peptides with PEG is often associated with an extension of circulatory half-life while cross-linking of proteins with glutaraldehyde, polyethyleneglycol diacrylate and formaldehyde is used for the preparation of hydrogels. Chemical modification of allergens for immunotherapy is often achieved by carbamylation with potassium cyanate.

Do not forget to describe the synthesis of said modified peptides in detail even if the methods are very well known in the art.

---

[29] Lundblad (2005).

[30] See Chap. 15 of Coligan et al. (2000) for more extensive methodology relating to chemical modification of proteins.

Nevertheless, this strategy will not always be successful and in some jurisdictions (see the chapter on US and China for more details) it might not help at all.

As an example, let us look at Korea. The following is an outtake of an office
action by the KIPO (Korea Intellectual Property Office).

The subject application cannot be allowed since Claims 1 to 18 thereof fail to
satisfy the description requirement under Article 42(4) (ii) of the Korean Patent
Law, as explained below:

(1) Although Claims 1 to 8 employ the expressions "a variant thereof that is at least 80%
    homologous to…," "a variant thereof," "peptide… comprises a continuous Stretch of
    amino acid," "peptide comprises the specific anchor amino acid-motif," "peptide
    consists essentially of," "modified and/or includes non-peptide bond," and "a fusion
    protein, in particular comprising N-terminal amino acids of…," the boundary of the
    claimed Peptides is unclear since they are not defined with a specific sequence.
    The subject application cannot be allowed since Claims 1 to 16 and 18 thereof fail to
    satisfy the description requirement under Article 42(4) (i) of the Korean Patent Law,
    as explained below:
(2) Although Claims 1 to 8 recite "a variant thereof that is at least 80% homologous to…,"
    "a variant thereof," "peptide being capable of stimulating CD4 and CD8," "peptide
    comprising a continuous Stretch of amino acids," "peptide comprising the specific
    anchor amino acid-motif," "peptide consisting or consisting essentially of…," "pep-
    tide being modified and/or including non-peptide bond," and "a fusion protein, in
    particular comprising N-terminal amino acids of…," the subject specification discloses
    only the immunogenic effects of peptides XXX-001 to XXX-005 corresponding to
    SEQ ID No. 1–5 in working examples. Since there is no evidence demonstrating that
    the claimed peptides indicated above have the similar immunogenic effects as peptides
    XXX-001 to XXX-005, the scope of these Claims is deemed much broader than what is
    supported by the specification. Further, Claims 9 to 16 and 18, depending on Claims 1
    to 8 directly or indirectly, have the same reason for rejection.
    The subject application cannot be allowed since the specification thereof fails to meet
    the description requirement under Article 42(3) of the Korean Patent Law, as explained
    below:
(3) Although Claims 1 to 8 recite "a variant thereof that is at least 80% homologous
    to…," "a variant thereof," "peptide being capable of stimulating CD4 and CD8,"
    "peptide comprising a continuous Stretch of amino acids," "peptide comprising the
    specific anchor amino acid-motif," "peptide consisting essentially of…," "peptide
    being modified and/or including non-peptide bond," and "a fusion protein, in par-
    ticular comprising N-terminal amino acids of…," the subject specification discloses
    only the immunogenic effects of peptides XXX-001 to XXX-005 corresponding to
    SEQ ID No. 1–5 in working examples. Therefore, those skilled in the art would not be
    able to confirm that the claimed peptides have the similar immunogenic effects as
    peptides XXX-001 to XXX-005, without undue burden of experimentation.

Basically, the KIPO is objecting on three different grounds and in that reasoning
shows how it leans on the USPTO and SIPO (China Intellcatual Property Office).

Objection (1) is due to allegedly not defined structures (although the variants
were explicitly cited in the application). This is the equivalent to a 112 rejection[31]

---

[31] Id 12.

issued by the USPTO. Objection (2) is trying to limit the scope of protection to the working examples. This is close to the Chinese way of limited protection (see the chapter on China for more details). Objection (2) refers to an allegedly undue burden of experimentation. The USPTO is always quick in issuing such rejections. It might not be restricted to the field of (therapeutic) peptides but we see a lot of times that the examiners do not know the art at all. Given the extensive art on how peptides work and how they can be modified without destroying the activity and given a description that already discloses variants and modifications to be employed and last but not least the speed of synthesis and the measurements needed to determine the immunogenic effects, the burden of experimentation for somebody with skill in the art (see artisan) is minor. Compare this with the billions of possible structures in small molecule applications (see above).

The objections are in line with the KIPO's Manual for Patent Examination in the Field of Biotechnology, (i) a polynucleotide or a DNA fragment described in the Claims should be specified by its nucleotide sequence or a polypeptide sequence encoded by such polynucleotide sequence; (ii) the polynucleotide or Polypeptide having at least X% homology to a certain polynucleotide or Polypeptide described in the claims should be supported by concrete examples of sequences having such homology in the specification (see also Patent Court Case Nos. 2007 Heo 289 dated May 29, 2008 and 2009 Heo 4261 dated May 14, 2010); and (iii) when variants of a polynucleotide or Polypeptide sequence resulting from Substitution, deletion or addition of one or more nucleotides or amino acids are claimed, the specific sites and kind of mutation should be specified and the concrete examples of such variants should be provided in the specification.

Article 42(4)(i) of the Korean Patent Law and Article 42(3) of the old Korean Patent Law stipulate that the Claims shall be supported by the specification; and that the specification shall state the purpose, constitution and effect of the invention in such a manner that it may easily be carried out by a person having ordinary skill in the art to which the invention pertains.

## 10.5 Decisions of the Board of the Appeal of the European Patent Office Concerning Pharmaceutical Active Peptides

Only one technical board decision[32] exists that concerns pharmaceutical active peptides. There is no special peptide- or biotech-related matter in the decision. In a nutshell, the granted patent did not enjoy the priority as claimed but nevertheless was novel.

---

[32] Technical Board Decision T 1414/05.

## 10.6 Examples of Specific Problems Related to Peptide-Related Inventions

### 10.6.1 Pharmaceutics

EP0434979 is an example for an early granted patent on pharmaceutical active patents being filed in 1990 and claiming priority to 1989. It is also an example for a granted Markush claim on (modified) peptides.

The first claims read as follows:

1. A peptide derivative of the formula X-N-$A_1$-$A_2$-$A_3$-$A_4$-$A_5$-$A_6$-$A_7$-$A_8$-$A_9$-$A_{10}$-$A_{11}$-$A_{12}$-$A_{13}$-$\Psi$-$A_{14}$-C-Y

   wherein

   X is an amino terminal residue selected from hydrogen, one or two alkyl groups in from 1–10 carbon atoms, one or two acyl groups from 2–10 carbon atoms, carbobenzyloxy or t-butyloxy carbonyl, unless the amino terminal amino acid is a cyclic derivative and thereby X is omitted,

   N is a bond, unless the amino terminal amino acid is a cyclic derivative and thereby N is omitted,

   $A_1$ is pGlu or a bond, $A_2$ is Gln or a bond, $A_3$ is Arg or a bond, $A_4$ is Leu or a bond, $A_5$ is Gly or a bond,

   $A_6$ is Asn, or D-Phe, $A_7$ is Gln or Eis, $A_8$ is Trp, $A_9$ is Ala, $A_{10}$ is Val, $A_{11}$ is Gly or D-Ala, $A_{12}$ is Eis,

   $A_{13}$ is Phe or Leu, $\Psi$ is a dipeptide determinant of $A_{13}$aa $\Psi$ $A_{14}$aa where $\Psi$ is [$CH_2S$], [$CH_2S(O)$],

   $\Psi$[$CH_2S(O)$]$^I$ or $\Psi$[$CH_2SO$]$^{II}$ where in $A_{13}$aa and $A_{14}$aa designate the substituent amino acids, $A_{14}$ is Leu,

   Nle or Met,

   C is a bond,

   Y is a carboxy terminal residue selected from OE, ($C_1$–$C_8$) alkoxyester, amino, mono or di ($C_1$–$C_4$) alkyl amide ($C_1$–$C_4$)alkylanine, ($C_1$–$C_4$)thioalkylether, or pharmaceutically acceptable salt thereof.

2. A peptide according to claim 1 wherein the peptide is pGlu-Gln-Arg-Leu-Gly-Asn-Gln-Trp-Ala-Val-Gly-His-Phe$\Psi$[$CH_2S$]Leu-$NH_2$, Ac-D-Phe-Gln-Trp-Ala-Val-D-Ala-His-Phe$\Psi$[$CH_2S$]Leu-$NH_2$,
   Asn-Gln-Trp-Ala-Val-Gly-HisPhe$\Psi$[$CH_2S$]Leu-$NH_2$, Asn-Gln-Trp-Ala-Val-Gly-His-Phe$\Psi$[$CH_2S(O)$]Leu-$NH_2$,pGlu-Gln-Trp-Ala-Val-Gly-His-Phe-$\Psi$[$CH_2S(O)$]Leu-$NH_2$,pGlu-Gln-Trp-Ala-Val-Gly-His-Phe$\Psi$[$CH_2S$]Nle-$NH_2$, pGlu-Gln-Trp-Ala-Val-Gly-His-Phe$\Psi$[$CH_2S(O)$]$^I$Leu-$NH_2$), pGlu-Gln-Trp-Ala-Val-Gly-His-Phe$\Psi$[$CH_2S$]Leu-$NH_2$ or pGlu-Gln-Trp-Ala-Val-Gly-His-Phe$\Psi$[$CH_2S(O)$]$^{II}$Leu-$NH_2$.

3. A process for preparing a peptide derivative according to claim 1 or 2 or a pharmaceutically acceptable salt thereof comprising the steps of:

   (a) using a resin with a suitably bound C-terminal protected dipeptide from the group $A_{13}\Psi A_{14}$, wherein $\Psi$ is [$CH_2S$] and $A_{13}$ and $A_{14}$ designates the substituent amino acids;

   (b) sequentially coupling the other alpha amino protected amino acids, $A_{12}$ through X, to achieve the protected amino acid sequence claimed;

   (c) removing said protecting groups;

(d) purifying the desired peptide, or optionally forming the $\Psi(CH_2S(O))]^I$, and $\Psi(CH_2SO)]^{II}$ isomers of said $\Psi[CH_2S]$ peptide and then purifying the desired isomeric peptide.

No SEQ IDs were given to any of the claimed or disclosed peptides. With 25 pages (including the translated claims) the disclosure is quite short. There are other examples like EP1210948 that are just three pages long. This is one of the most obvious differences to applications filed nowadays. Just the description of the underlying technique is in many cases much longer.

While the short disclosure of this example was enough to get a patent granted, short disclosures often lead to "un-claimable" subject matter. One of the many examples is EP0811013. The last office action stipulated the following:

> [...] sequence needs support in the application documents as originally filed. This is not the case since the new concensus sequence now claimed is not disclosed per se in the description or Claims as originally filed. Since the requirements of Article 123 (2) EPC are not fulfilled Claim 1 will not be examined. [...]
> The subject-matter of Claim 5, however, cannot be found in the application documents as originally filed. The applicants should note that the examining division has to take a strict position when the requirements of Article 123 (2) EPC are concerned and has no discretion whatsoever to allow amendments that might be obvious modifications or embodiments of what has been disclosed in the original text of the specification. ...

The longer the better? Far from that! A few years back applicants often tried every single possibility to describe the invention so that nobody would understand what they really disclosed. They ran in exactly the problems mentioned above.

It was taught to beginners that you should cover the "real" invention in as much dust as possible, so that it would be difficult for others to make use of the invention. So the applications got longer and longer, publication and attorney fees were growing, but what is achieved?

Let us have a look at a relatively new example taken from US20070185025[33]

> ...The invention provides, an isolated peptide comprising the consecutive amino acid sequence of any one of SEQ ID NOS: 1-84, 108-376, wherein the total length of the peptide is less than about 66, 64, 62, 60, 58, 56, 54, 52, 50, 48, 46, 44, 42, 40, 38, 36, 34, 32, 30, 28, 26, 24, 22, 20, 18 amino acids and wherein the peptide has immunosuppressive activity. ...

Even at the beginning, not only about 350 different peptides are "disclosed", but also little snips with total lengths between 18 and 66 amino acids. In all jurisdictions such claim wording leads to problems with written description (e.g., what is the actual peptide?) and undue burden on the artisan, who actually has to synthesize and test all possible peptides for immunosuppressive activity to find a candidate or to determine whether his compound falls under the description (claims). Another example from US20070185025 reads as follows:

---

[33] This application is assigned to the Columbia University New York, but similar problems exist with other university applications.

...The invention provides, an isolated therapeutic peptide comprising: NRXX(X1)DXL(X2)X(R)XXXXC sequence motif, wherein X is any amino acid, (X1) is leucine, or isoleucine, (X2) is leucine, isoleucine, or Phenylalanine, (R) is arginine or lysine, wherein the peptide length is from about 16 amino acids to about 66, 64, 62, 60, 58, 56, 54, 52, 50, 48, 46, 44, 42, 40, 38, 36, 34, 32, 30, 28! 26, 24, 22, 20, 18 amino acids...

As discussed for the applicability of Markush claims to "BioTech" claims, the applicant tries to cover a motif but at the same time broadens the motif to lengths between 16 and 66 amino acids. Again the problems with written description are obvious. And furthermore, the peptides according to this description do not have to show any immunosuppressive activity. Yet another example from US20070185025 reads as follows:

...The invention provides an isolated peptide having amino acid sequence of SEQ ID NO: 1 or having an amino acid sequence which is at least about 70%, 72%, 74%, 76%. 78%, 80%, 82%, 84%, 86%, 88%, 90%, 91%, 92%, 93%, 94%, 95%, 96%, 97%, 98%, 99%, 99.5% identical to the amino acid sequence set forth in SEQ ID NO: 1. The invention also provides an isolated peptide having amino acid sequence of SEQ ID NO: 2 or having an amino acid sequence which is at least about 70%, 72%, 74%, 76%, 78%, 80%, 82%, 84%, 86%, 88%, 90%, 91%, 92%, 93%, 94%, 95%, 96%, 97%, 98%, 99%, 99.5% identical to the amino acid sequence set forth in SEQ ID NO:2 [this goes on for the other SEQ IDs]...

Today, applicants are trying to broaden the disclosure on the peptides by disclosing "identity rows". If you do not write it, it is not in there, so the percentages for every SEQ ID NO are put down. Besides the already mentioned problems with written description, applicants also forgot to define "identical". Beware of attorneys trying to talk you into such things. It almost looks like the applicant does not know what the actual invention is. Still another example from US20070185025 reads as follows:

... The invention provides an isolated peptide selected from the group consisting of sequences with SEQ ID NOS. 1 to 84, or any other peptide of the invention. The invention also provides an isolated peptide comprising from about 9, 10, 11, 12, 13, 14, 15 to about 11, 12, 13, 14, 15, 17, 19, 21 23, 25, 27, 29, 31, 33, 35, 37, 39, 41, 43, 45, 50, 55, 60, 65, 70, 75 consecutive amino acids from a Polypeptide, for example but not limited to a filoviral glycoprotein Polypeptide, wherein at least a portion of the amino acid sequence can form a coiled-coil secondary structure. In certain aspects, the consecutive amino acids comprise an amino acid sequence motif RXXXD wherein X can be any amino acid; an amino acid sequence motif comprising two arginines separated from each other by at least eight amino acids, such as RXXXDXXXXD, wherein the RXXXD motif is between the two arginines. In other aspects, the consecutive amino acids can form a secondary structure similar or identical to the secondary structure of the carboxy terminus domain of the retroviral env protein. ...

In this part the disclosure names to proteins from which the CAN be obtained from. Further another motif is disclosed that—again—broadens the motif disclosed *supra*, as X1 is no substituted by X. Another example from US20070185025 reads as follows:

... In one aspect the invention provides a peptide, which may be at least 75% identical to a peptide of any one of SEQ ID NOs: 1-84, 108-376, or any other peptide of the invention.

In one embodiment the homology can be between 75% and 79.99%. In another embodiment the homology can be between 80% and 84.99%. In another embodiment the homology can be between 85% and 89.99%. In another embodiment the homology can be between 90% and 94.99%. In another embodiment the homology can be between 95% and 99.99%. ...

The applicant started with "identical" and gave a quite broad range of possible "identities". As you remember "identical" was not defined. So why not try "homology" and a range that goes up to startling 99.99%? You guessed it! Homology is also not defined. In the application a lot of the sequences disclosed are around 17 amino acids in length. So everything with a homology between roughly 95 and 100% must be identical. Another example from US20070185025 reads as follows:

... In another embodiment of the invention, the peptides may also be prepared and stored in a salt form. Various salt forms of the peptides may also be formed or interchanged by any of the various methods known in the art, e.g., by using various ion exchange chromatography methods. Cationic counter ions that may be used in the compositions include, but are not limited to, amines, such as ammonium ions, metal ions, especially monovalent, divalent, or trivalent ions of alkali metals including sodium, potassium, lithium, cesium; alkaline earth metals including calcium, magnesium, barium; transition metals such as iron, manganese, zinc, cadmium, molybdenum; other metals like aluminum; and possible combinations of these. Anionic counter ions that may be used in the compositions described below include chloride, fluoride, acetate, trifluoroacetate, phosphate, sulfate, carbonate, citrate, ascorbate, sorbate, glutarate, ketoglutarate, and possible combinations of these. Trifluoroacetate salts of peptide compounds described here are typically formed during purification in trifluoroacetic acid buffers using high-performance liquid chromatography (HPLC). Although usually not suited for in vivo use, trifluoroacetate salt forms of the peptides described in this invention may be conveniently used in various in vitro cell culture studies, assays or tests of activity or efficacy of a peptide compound of interest. The peptide may then be converted from the trifluoroacetate salt by ion exchange methods or synthesized as a salt form that is acceptable for pharmaceutical or dietary Supplement compositions. ...

It seems likely that the attorney asked the applicant what salts there possibly might be. But for use in animals (not to talk about humans) some of the disclosed ions such as cesium or barium are usually not suited for use in vivo.

The problems patent authorities have with the state of the art in high-tech areas like active peptides were briefly touched. However, not only patent authorities have problems to keep pace with speed of development in biotec. Take for example WO2004110472, which has the following passage:

Various methods of protein purification may be employed and such methods are known in the art and described, for example, in Deutscher, Methods in Enzymology 182: 83-9 (1990) and Scopes, Protein Purification: Principles and Practice, Springer-Verlag, NY (1982). The purification step(s) selected will depend on the nature of the production process used and the particular heterologous fusion protein produced. "or " A suitable selection gene for use in yeast is the trpl gene present in the yeast plasmid YRp7 [Stinchcomb, et al., Nature 282(5734): 39-43 (1979); Kingsman, et al., Gene 7(2): 141-52 (1979); Tschumper, et al., Gene 10(2): 157-66 (1980)]. The trpl gene provides a selection marker for a mutant strain of yeast lacking the ability to grow in tryptophan, for example, ATCC No. 44076 or PEPC1 [Jones, Genetics 85: 23-33 (1977)].

Very often, applicants receive office actions citing prior art which is very oldart that old and claiming the "unpredictability" of the art in view of that art, the quality of those rejections is sometimes pretty disturbing.

Attorneys as well as companies should also beware of "copy paste" applications, with parts on the underlying technologies being taken year after year and application after application. This will work for definitions (a rose is a rose, a carbon atom is a carbon atom) and maybe the very foundations of the art. It is *okay* to lay the ground by mentioning the beginning of a development, like "the first clinical trials with XYZ started in 1892", or "the general concept of density was developed by Archimedes of Syracuse". But in a rapidly changing field as Genetics or BioTech techniques which have been high-end a decade ago may have become the laboratory standard today.

Before copy-and pasting prior art discussions from old applications (because one always cited it and one have it in the right format), one should think twice whether that art is really needed in the disclosure and whether the art is still applicable. In the US this might even lead to a "112 rejection"[34] as the product might not be synthesized according to the methods cited by the applicant (or if one incorporated something by reference).

There is one general problem related to cells that may be claimed in context with peptides. This is not the place to discuss the rulings on stem cells and especially human embryonic stem cells in the various jurisdictions. But in a lot of applications either cells to produce the disclosed peptide(s) or cells that will express the peptides in vivo (given back to a patient in need) are claimed. To be on the safe side those cells should always be described in detail or at least be named. I have seen many objections around the world against claimed cells for ethical reasons or trying to patent a human being based on such a cell.

Will disclosing certain cell types for depending claims protect you from all "cell based" objections? Certainly not... Maybe the examiner will not know what—for example—Awells cells are, but that can be easily explained. Even a disclaimer excluding embryonic stem cells would help, as long as it is in the description. It is true that sometimes such a disclaimer can be put in afterwards, even if it is not in the original disclosure, but it is much easier having the description to the point.

### 10.6.2 Vaccines

Most of what was said and could be seen in the examples for "pharmaceutics" also applies to application/patents on peptide-based vaccines.

There are two trends one can see in filings concerning peptide vaccines. These trends have their roots equally in the technique and the proceedings before the PTOs.

---

[34] Id 12.

When the first applications concerning peptide-based vaccines were filed a lot of the applications disclosed hundreds or even thousands of peptide sequences. The applicants simply disclosed all peptides they found in their research (most times cell lines). The experiments to check for the immunogenicity of the found peptides were very slow. This is why most of the applications show any data for just one or two of the peptides and it was simply stated that the other peptides would show comparable functionalities. For example US6413517 disclosed 274 sequences, US6037135 discloses 1254 sequences. There were a lot of applications filed until about 2006 with SEQ ID NOs in that range (or even more). Even today some applications disclose vast numbers, but this has become rarer. This is due to the fact that the determination of the immunogenicity of the found peptides and as the first step the synthesis of the peptides has become easier and therefore instable, insoluble, or peptides without immunogenicity are kept out of the application. Another downside of including vast numbers of peptides in an application will be discussed in detail in the chapter about the situation in China. In a nutshell, in China and in some other jurisdictions there will not be a patent issued for any subject matter without any proof in the filed disclosure that the claimed subject matter (in this case peptides) exhibits the claimed properties.

In some cases even the applicant seems to get a little lost if the numbers of disclosed peptides get too high. In US20080207497 no less than 1533 peptides are disclosed by sequence ID (not including variations and the like). But some of the sequences were already disclosed by the same applicant already 6 years earlier in WO200246416 and the same sequences were disclosed inter alia in WO2008070047 and WO2008088583. Unfortunately, for the applicant the first disclosure renders the disclosed peptides not novel and obvious.

The other trend follows the same path. Some early applications not only cited hundreds of peptides but also calculated the peptides. Take for example WO2004052917 and WO2002061435 and the related patent families. Some of the applications disclose several hundreds of "immunogenic" peptides. Those "immunogenic" peptides are identified using algorithms[35]:

> The algorithms are mathematical procedures that produce a score which enables the selection of immunogenic peptides. Typically one uses the algorithmic score with a "binding threshold" to enable selection of peptides that have a high probability of binding at a certain affinity and will in turn be immunogenic. The algorithm is based upon either the effects on MHC binding of a particular amino acid at a particular position of a peptide or the effects on binding of a particular substitution in a motif containing peptide.

Some of the algorithms employ the database "SYFPHEITI" as a reference and the "SYFPEITHI Score" as the determinant of the immunogenicity. This method is commonly described as reverse immunology, the procedure to predict and identify immunogenic peptides from the sequence of a gene product of interest. It has been postulated to be a particularly efficient, high-throughput approach for tumor antigen discovery. Viatte et al. conclude "that the overall low sensitivity

---

[35] Definition taken from WO2002020616.

and yield of every prediction step often requires a compensatory up-scaling of the initial number of candidate sequences to be screened, rendering reverse immunology an unexpectedly complex approach.[36]"

The problem for applications based on the reverse immunology is obvious. Again, problems will arise in jurisdictions where the properties have to be proven in the disclosure, if there are unstable or unsoluble peptides in the disclosure and this can be shown written description and undue experimentation as well as enablement will be more than problematic. While in the first examples the peptides were "isolated" or "found" in a living system, the pure theoretical disclosure of most peptides changes the situation for the applicant dramatically.

We found one example where such a situation was mentioned in observations of third parties before the EPO. In this communication (the application is EP1931377) the lack of enablement is one of the many points raised by the third party.

The application in dispute concerns CD4 + T survivin epitopes and their vaccine and diagnostic uses. In brief, Claims 1 to 37 of the application in dispute are directed at a peptide from the isoform alpha of survivin for vaccination, the diagnosis, and treatment and monitoring of cancer, wherein the said peptide can be selected from the group consisting of:

(a) the peptides of 13–18 consecutive amino acids located between positions 17 and 34 of the alpha-isoform of survivin,
(b) the peptides of 13–30 consecutive amino acids located between positions 84 and 113 of the alpha-isoform of survivin,
(c) the peptides of 13–21 consecutive amino acids located between positions 222 and 142 of the alpha-isoform of survivin, and
(d) the variants of the peptides defined in (a), (b) or (c),said peptides in (a), (b) or (c), or variants in (d), having a binding activity with respect to at least one HLA II molecule predominant in the Caucasian population, of less than 1,000 nM,

and being capable of inducing survivin-specific CD4 + T lymphocytes.

Explicitly mentioned—amongst others—are the peptides 17–31 (SEQ ID No. 5), 19–33 (SEQ ID No. 6), 20–34 (SEQ ID No. 7), 84–98 (SEQ ID No. 17), 90–104 (SEQ ID No. 19), 91–105 (SEQ ID No. 20), 93–107 (SEQ ID No. 21), and 96–110 (SEQ ID No. 23).

The subject matter of claim 1 does not comply with Article 83 EPC. The invention is disclosed in a way that it can be carried out by a person skilled in the art in its entirety.

The peptide described by the invention located between amino acids "94–108", and therefore included in claim 1(b), is insoluble, as the inventors admit themselves in "tableau IV" on page 30 within the specification of the application. Insoluble peptides cannot be used for the stipulated purposes of claim 1, and

---

[36] Viatte et al. (2006).

further, since the solubility is a requirement for the induction of a survivin specific CD4 + T-lymphocyte response, do not have the alleged technical effect as required by claim 1. Thus, the person of skill in the art is not sufficiently enabled to carry out the invention over the whole range as claimed.

Although it might be rare that proof or indications that some of the claimed and disclosed peptide are not suitable can be found in the disclosure itself the risk is obviously very high that some peptides that will not work can be found in peptide laundry lists as described.

One extreme example for patent claims filed in the early days of research on peptide vaccines can be found in WO0006723:

1. A peptide derived from a protein selected from the group consisting of Uroplakin (UP), Prostate specific antigen (PSA), Prostate specific membrane antigen (PSMA), Prostate acid Phosphatase (PAP), Lactadherin (BA-46), Mucin (MUC1) and Teratocarcinoma-derived growth factor (CRIPTO-1), the peptide comprising 8 to 10 amino acid residues, of which a second residue from an amino terminal of the peptide and an end residue at a carboxy terminal of the peptide are hydrophobic or hydrophilic natural or non-natural amino acid residues.

# 11 The Situation in the USA[37]

This section is intended to give a broad overview of areas of law that are particularly relevant to the patenting of therapeutic peptides as a product and in compositions. As will be well understood, every technical requirement for patentability has its own vagaries with respect to particular technology areas. As such, a comprehensive discussion will be unwieldy. Rather than do that, this focuses on areas that have evolved as hot areas before examiners.

The scope of patentable subject matter will be discussed, as it has been the focus of much wrangling and debate in the courts. In particular, the Federal Circuit opinion in the *Myriad* case will be discussed, including its potential implication on the patentability of naturally produced peptides.

In addition, current USPTO policy with respect to the written description requirement and Markush claims are addressed, as examiners are being pushed to apply these doctrines to narrow claims in lieu of prior art.

Finally, the legal requirements for "obvious to try" rationale to establish a prima facie case of obviousness will be addressed, as the *KSR* decision has emboldened examiners to push the limits of this doctrine.

The recently decided case of Mayo vs Prometheus[38] will be discussed in depth in a following book of this series.

---

[37] By Bryan W. Jones of Baker, Donelson, Bearman, Caldwell & Berkowitz, PC.

[38] *Mayo Collaboration Services vs Prometheus Labs., Inc.,* Slip. Op. Dkt. No. 10–1150, 566 (2012).

## 11.1 Patentable Subject Matter

In the United States, the scope of patent-eligible subject matter is quite broad, extending to "any new and useful process, machine, manufacture, or composition of matter, or any new and useful improvement thereof...[39]" It is not unlimited, however. Courts have long read 35 U.S.C. § 101 as excluding "the laws of nature, physical phenomena, and abstract ideas" from patent protection.[40] This has been applied to exclude certain categories of subject matter from patent eligibility, including: (1) bare mathematical formulae[41]; (2) methods in which "the only novel element. . . is a mathematical formula[42]"; (3) "propagating electromagnetic signals[43]"; and (4) mixtures of naturally occurring microorganisms. Additionally, Congress recently codified USPTO policy holding that claims covering human organisms are patent-ineligible.[44]

USPTO policy considers naturally occurring products to be patent eligible "when isolated from their natural state and purified or when synthesized in a laboratory from chemical starting materials.[45]" As explained by the USPTO:

> When Congress enacted the patent statutes, it specifically authorized issuing a patent to a person who "invents or discovers" a new and useful composition of matter, among other things. The pertinent statute is 35 U.S.C. § 101, which reads: "Whoever invents or discovers any new and useful process, machine, manufacture, or composition of matter, or any new and useful improvement thereof, may obtain a patent therefor, subject to the conditions and requirements of this title." Thus, an inventor's discovery of a gene can be the basis for a patent on the genetic composition isolated from its natural state and processed through purifying steps that separate the gene from other molecules naturally associated with it.
>
> If a patent application discloses only nucleic acid molecular structure for a newly discovered gene, and no utility for the claimed isolated gene, the claimed invention is not patentable. But when the inventor also discloses how to use the purified gene isolated from its natural state, the application satisfies the "utility" requirement. That is, where the application discloses a specific, substantial, and credible utility for the claimed isolated and purified gene, the isolated and purified gene composition may be patentable.[46]

This policy finds support in cases from the US District Court for the Southern District of New York ("SDNY") and the US Court of Customs and Patent Appeals ("CCPA"), each of which found isolated and/or purified biological products (adrenaline and prostaglandin, respectively) to satisfy 35 U.S.C. § 101.[47] The

---

[39] 35 U.S.C. § 101.

[40] Diamond v. Chakrabarty, 447 U.S. 303, 309, 100 S.Ct. 2204 (1980).

[41] Gottschalk v. Benson, 409 U.S. 63, 71–73, 93 S.Ct. 253 (1972).

[42] Parker v. Flook, 437 U.S. 584, 98 S.Ct. 2522 (1978).

[43] *In re Nuijten*, 500 F.3d 1346, 1355–56 (Fed. Cir. 2007).

[44] See Smith-Leahy America Invents Act, Pub. L. 112-29 at § 33(a) (Sep. 16, 2011).

[45] Utility Examination Guidelines, 66 Fed. Reg. 1092 at 1093 (Jan. 5, 2001).

[46] Id.

[47] See id. at (citing *Parke-Davis & Co.* v. *H. K. Mulford Co.*, 189 F.2d, 103 (S.D.N.Y. 1911); and *In re Bergstrom*, 427 F.2d 1394, 1401 (CCPA 1970).

USPTO has explicitly adopted the same rationale with respect to peptides, allowing numerous patents directed to isolated polypeptides to issue.[48] Thus, so long as the specification discloses a "specific, substantial, and credible utility" for a peptide, the USPTO allows it to be claimed in its isolated, purified, or synthetic form.[49]

However, this "isolation" rationale has come under fire in the courts recently. In 2009, a United States District Court held that a patent directed to isolated BRCA genes, isolated cDNA encoding BRCA proteins, and fragments thereof fail to satisfy 35 U.S.C. § 101.[50] According to the court, an isolated product of nature does not satisfy 35 U.S.C. § 101 unless it is "markedly different" in the isolated form as compared to its naturally occurring form.[51] Isolated nucleic acids do not satisfy this test because they are simply "physical embodiments of information," the content of which is identical to that of the naturally occurring molecules. The CAFC ultimately reversed, holding that each category of claims was patent eligible. However, a majority rationale could not be reached with respect to the cDNA, while the gene claims and fragment claims were allowed over a dissenting opinion by Judge Bryson and a less-than enthusiastic concurrence by Judge Moore.[52]

Judge Lourie, finding in favor of all three categories, reasoned that "it is the distinctive nature of DNA molecules as isolated compositions of matter that determines their patent eligibility rather than their physiological use or benefit.[53]" Because isolated nucleic acids are chemically distinct from the naturally occurring nucleic acids, the mere fact that they share a common coding sequence is thus inconsequential. Judge Lourie further distinguished "isolated DNA" from "purified" DNA, reasoning that "isolating" nucleic acids involves both removing DNA

---

[48] See, e.g., US Pat. No. 7,470,669 (filed Sep. 27, 2005).

[49] See M.P.E.P. § 2107(II)(A).

[50] See *Association for Molecular Pathology v. USPTO*, Dkt No. 2010-1406, — F.3d ——, 2011 WL 3211513, 99 U.S.P.Q.2d 1398 (Fed. Cir. Jul. 29, 2011).

[51] See id. at *2.

[52] Id. at *29. Judge Moore only held such claims to be patentable because the USPTO has a long history of allowing such claims. As she stated: If I were deciding this case on a blank canvas, I might conclude that an isolated DNA sequence that includes most or all of a gene is not patentable subject matter. Despite the literal chemical difference, the isolated full length gene does not clearly [*98] have a new utility and appears to simply serve the same ends devised by nature, namely to act as a gene encoding a protein sequence. This case, however, comes to us with a substantial historical background. Congress has, for centuries, authorized an expansive scope of patentable subject matter. Likewise, the United States Patent Office has allowed patents on isolated DNA sequences for decades, and, more generally, has allowed patents on purified natural products for centuries. There are now thousands of patents with claims to isolated DNA, and some unknown (but certainly large) number of patents to purified natural products or fragments thereof.n4 As I explain below, I believe we must be particularly wary of expanding the judicial exception to patentable subject matter where both settled expectations and extensive property rights are involved. Combined with my belief that we should defer to Congress, these settled expectations tip the scale in favor of patentability.
Id.

[53] Id. at *18.

from its native environment, and manipulating it chemically "so as to produce a molecule that is markedly different from that which exists in the body," whereas "purification makes pure what was the same material, but was previously impure.[54]" Unfortunately, Judge Lourie does not explain whether this distinction matters and, if so, how. Finally, Supreme Court precedence requires that deference be given to the PTO's history of granting such patents, so as to avoid "disrupting the settled expectations of the inventing community.[55]" In this case, the PTO had granted over 40,000 DNA-related patents over almost 30 years.[56] As such, any change in policy should come from Congress, not the courts.[57]

Judge Moore concurred in the result, holding that isolated nucleic acids satisfy 35 U.S.C. § 101. However, she set forth a different rationale from that applied by Judge Lourie. Judge Moore separated isolated nucleic acids into essentially two categories: (1) those having a contiguous sequence that is different from naturally occurring nucleic acids (ex. covering cDNA); and (2) those having a contiguous sequence that is identical to a contiguous sequence of a naturally-occurring nucleic acid (ex. covering probes and genes).[58] With respect to the first category, Moore reasoned that these clearly satisfy 35 U.S.C. § 101, as they are not naturally occurring products.[59] With respect to the second category, however, the patentability lies not only in differences in chemical structure, but also in expanded scope of utility.[60] Under this rationale, Moore would have held that claims covering shorter sequences, such as probes or primers, satisfy 35 U.S.C. § 101, whereas claims covering full-length genes would not.[61] However, given the long-standing practice of the USPTO granting applications directed to isolated nucleic acids, Moore concurred in the result and found all claims to satisfy 35 U.S.C. § 101.[62]

Judge Bryson dissented, arguing that patentability must be analyzed by focusing on: (1) "similarity in structure between what is claimed and what is found in nature"; and (2) "similarity in utility between what is claimed and what is found in nature.[63]" Claims encompassing full-length genes are not patentable subject matter because "the only material change made to those genes from their natural state is the change that is necessarily incidental to the extraction of the genes from the environment in which they are found in nature.[64]" Contrasting genes with the

---

[54] Id.

[55] Id. at *19. Citing *Festo Corp. v. Shoketsu Kinzoku Kogyo Kabushiki Co.*, 535 U.S. 722, 739 (2002).

[56] Id. at *20.

[57] Id. at *19.

[58] Id. at *28.

[59] Id. at *28–29.

[60] Id. at *29.

[61] Id.

[62] Id. at *31.

[63] Id. at *42.

[64] Id. at *38.

example of a baseball bat made from an ash tree, he reasoned that a baseball bat is patent eligible because "the process of 'extracting' the bat necessarily changes the nature, form, and use of the ash tree and thus results in a manmade manufacture, not a naturally occurring product, whereas genes are not patent eligible because "nature has defined genes as independent entities by virtue of their capacity for protein synthesis. ...[65]" Bryson further distinguished "purification" cases on the basis that purification must result in a "marked change in functionality" in order to be patentable subject matter, which allegedly is not the case with isolated genes.[66] Moreover, with respect to claims drawn to shorter sequences, he argues that such claims are quite broad, reading on probes and primers, but also introns, single exons, and sequences expected to be found in other nucleic acids that do not encode BRCA.[67]

The outcome of course has implications for therapeutic peptides, as well as other naturally occurring biological molecules. The one point that all Judges seemed to agree on is that an isolated compound satisfies 35 U.S.C. § 101 when it either is a completely new sequence (ex. a cDNA) or when it has a utility in its isolated state that it does not have in its natural state. The first circumstance makes sense, as such claims relate to purely man-made products. The second circumstance also makes sense as a logical extension of the "purification" cases. The question remains, however, as to how "new" the utility must be for isolated naturally occurring molecules. For example, how would the Courts treat a claim directed to an isolated gene encoding a polypeptide that has a disclosed therapeutic use. True, such a nucleic acid will perform the same discrete task as the naturally occurring gene, namely, drive expression of the polypeptide. However, the broader utility in the context of such an invention is to form a transformed host cell for the constitutive expression of the peptide, a task that could not be performed by the nucleic acid in its natural state.

A petition for certiorari is still pending before the Supreme Court, so it may be some time before a definitive opinion is reached.

## 11.2  35 U.S.C. 112, First Paragraph: Written Description and Enablement

One quirk of biotechnological applications is that they are far more likely than applications in other art units to receive a rejection under 35 U.S.C. § 112.[68] Indeed, rejections under 35 U.S.C. § 112 are more common than rejections based

---

[65] Id.

[66] Id. at *42.

[67] Id. at *43.

[68] Cotropia et al. (2010). The next closest art unit is 1700 (chemicals), in which 56% of cases receive a "112 rejection".

on art in these cases.[69] As such, it is important to know some of the hot button issues that give rise to many of these rejections.

## 11.2.1 Written Description

Unlike many places throughout the world, the USPTO does not require a claim limitation to be supported by an *ipsis verbis* recitation in the specification.[70] Rather, all that is required is a sufficiently detailed description such that "possession" of the claimed subject matter is demonstrated by the specification at the time of filing.[71] This in many respects is a flexible standard that is relatively easy to satisfy. However, the USPTO has recently urged examiners to apply it more frequently to narrow unduly broad generic claims.[72] This has proven to be a major hurdle to overcome in obtaining a reasonable claim directed to biological molecules.

In order to adequately describe a generic claim, the specification must set forth either a "representative number" of species within the scope of the claim or sufficient characteristics common to all members of the genus that comprise a substantial portion of the claimed molecule. Federal circuit precedent makes clear that a detailed description of each species falling within the genus is not required.[73] Indeed, there are cases in which working examples within the scope of the claim are not even required.[74] In the typical case, however, the claim should recite some structural features common to all members of the genus, although the description need not be greatly detailed in all cases.

For example, USPTO training materials indicate that a claim directed to "a peptide comprising SEQ ID NO: X" normally is supported by a description of a single working embodiment directed to a peptide consisting of that sequence.[75] The rationale for this policy is clear: if the peptide consisting of the sequence possesses a function, the addition of sequences to either end of the peptide is unlikely to affect that function of the overall peptide.[76]

Recitation of a peptide "having X% sequence identity with SEQ ID NO: X" should likewise be supported by a disclosure of a peptide consisting of SEQ ID

---

[69] Id. (showing that, in Art Unit 1600, 76% of all cases receive a rejection under 35 U.S.C. 112, whereas 67% of cases receive a rejection under 35 U.S.C. 102 or 103).

[70] M.P.E.P. § 2163 (*citing Martin v. Johnson*, 454 F.2d 746, 751, 172 USPQ 391, 395 (CCPA 1972)).

[71] Id. (*citing Capon v. Eshhar*, 418 F.3d 1349, 1357, 76 USPQ2d 1078, 1084 (Fed. Cir. 2005)).

[72] David Kappos' Public Blog, "Written Description–Little Used Perhaps, But Extremely Useful to Ensure Claims are Appropriately Scoped" (May 5, 2010), *available at* http://www.uspto.gov/blog/director/entry/written_description_little_used_perhaps (last accessed Jan. 16, 2012).

[73] *Falkner v. Inglis*, 448 F.3d 1357, 1368 (Fed. Cir. 2006).

[74] Id.

[75] Written Description Training Materials, Revision 1 at Example 9, page 31 (Mar. 25, 2008).

[76] Id.

NO: X.[77] The rationale for this policy is that a person of ordinary skill in the art would be expected to understand the extent to which a given peptide may be modified without significant loss of tertiary structure.[78]

The situation changes once functional limitations are added to the mix, however. The USPTO requires the applicant to establish a relationship between the recited function and some structure common to all members of the genus.[79] For example, a claim that simply recites "a peptide having X% identity to SEQ ID NO: Y having the function Z" will generally be rejected for lack of written description support absent some showing of the structure necessary for the peptide to possess the recited function.[80] This presents a rather odd situation where a relatively broad generic claim satisfies the patent act, whereas a narrower subgenus of that claim does not.

However, a recitation of sequences necessary for a specific function in the claim is not required if a representative number of sequences sufficient for the recited function are disclosed in the specification. This principle is illustrated nicely by the case of *Ex parte Joo-Eun Bae*.[81] Claim 1 is illustrative of the claims at issue therein:

1. An isolated leukemic antigen consisting of a fragment of CD19 (SEQ ID NO: 13) or a variant thereof having one, two, or three conservative or non-conservative amino acid substitutions that is capable of stimulating a cytotoxic T-lymphocyte reaction against a cell expressing CD19, wherein the fragment is
   (a) 11 to 80 amino acids in length or
   (b) 9 to 80 amino acids in length and comprises the amino acid sequence RLLFFLLFL (SEQ ID NO: 1), TLAYLIFCL (SEQ ID NO: 2), LLFLTPMEV (SEQ ID NO: 3), KLMSPKLYV (SEQ ID NO: 4), or LLFFLLFLV (SEQ ID NO: 5).

As can be seen, the peptides are claimed according to:

- *structure* ("consisting of a fragment of CD19 (SEQ ID NO: 13)··· (a) 11–80 amino acids in length or (b) 9–80 amino acids in length);

---

[77] Id. at Example 10, Claim 2, pages 34–35.

[78] Id. at Examples 10 and 11, page 34–37. As explained by the materials:

In this example, there is no disclosure relating similarity of structure to conservation of function. General knowledge in the art included the knowledge that some amino acid variations are tolerated without losing a protein's tertiary structure. The results of amino acid substitutions have been studied so extensively that amino acids are grouped in so-called "exchange groups" of similar properties because substituting within the exchange group is expected to conserve the overall structure. For example, the expectation from replacing leucine with isoleucine would be that the protein would likely retain its tertiary structure. On the other hand, when non-exchange group members are substituted, e.g., prolinefor tryptophan, the expectation would be that the substitution would not likely conserve the protein's tertiary structure. Given what is known in the art about the likely outcome of substitutions on structure, those in the art would have likely expected the applicant to have been in possession of a genus of proteins having a tertiary structure similar to SEQ ID NO: 2 although the claim is not so limited.

Id. at 38–39.

[79] Id. at 33–42.

[80] Id.

[81] *Ex parte Joo-Eun Bae*, Appeal 2009-013469, Application 10/884,862, Decision on Appeal (BPAI Apr. 5, 2010).

- *percent identity* ("a variant thereof having one, two, or three conservative or non-conservative amino acid substitutions"); and
- *function* ("capable of stimulating a cytotoxic T-lymphocyte reaction against a cell expressing CD 19").

The specification disclosed 180 nonamers of SEQ ID NO: 13 predicted by an MHC-binding algorithm to be capable of binding to a specific MHC class I allele and provided in vitro MHC binding data for five of the specific peptides identified.[82] Additionally, evidence was presented showing that the algorithm "predicts peptide/MHC interactions well.[83]" The BPAI found this evidence to be persuasive with respect to written description, arguing that "this evidence supports the conclusion that peptides drawn from multiple different regions of CD19 will be capable of stimulating the requires (*sic*) cytotoxic T-lymphocyte reaction.[84]" What will be considered a representative number of species will necessarily vary according on a case-by-case basis. Nonetheless, the burden always rests with the examiner to advance reasons for why the claim lacks description.

In contrast, claims that are defined solely according to function will generally require a much more detailed description in the specification. That being said, a working example within the scope of the claim is not always necessary, particularly where the missing descriptive matter was well know in the prior art.[85] In *Falkner v. Inglis*, the Federal Circuit held that a claim directed to a vaccine comprising a poxvirus having a genome in which a gene necessary for replication of the virus is inactivated, despite no examples using poxvirus and no disclosure of genes that would satisfy the claim.[86] Critically, the undisputed facts showed that poxvirus genes necessary for replication in a whole cell were well known at the time of filing, and a person of ordinary skill in the art would understand how to inactivate them. In so holding, the Court emphasized three points: (1) examples are not necessary to support the adequacy of a written description (2) the written description standard may be met (as it is here) even where actual reduction to practice of an invention is absent; and (3) there is no per se rule that an adequate written description of an invention that involves a biological macromolecule must contain a recitation of known structure.


## 11.2.2 Enablement

Pursuant to 35 U.S.C. 112, first paragraph, the specification must "enable any person skilled in the art... to make and use the claimed invention. ..." The ultimate question in determining enablement is whether the specification would

---

[82] Id.

[83] Id.

[84] Id.

[85] *Falkner v. Inglis*, 448 F.3d 1357, 1368 (Fed. Cir. 2006).

[86] Id.

permit a person of ordinary skill in the art to make and use the claimed subject matter without "undue experimentation". The Federal Circuit has set forth seven factors (referred to as "the Wands factors") which must be considered in determining whether undue experimentation is required:

(A) The breadth of the claims;
(B) The nature of the invention;
(C) The state of the prior art;
(D) The level of predictability in the art;
(E) The amount of direction provided by the inventor;
(F) The existence of working examples; and
(G) The quantity of experimentation needed to make or use the invention based on the content of the disclosure.[87]

Upon consideration of these factual elements, the examiner must determine whether the degree of enablement is reasonably commensurate to the scope of the claim at issue. In other words, the ultimate question is whether a person of skill in the art would have been able to apply the teachings of the specification to practice the full scope of the claims with nothing more than routine experimentation. As can be seen, this is a fact-driven determination and must be evaluated on a case-by-case basis.

In the therapeutic/pharmaceutical context, the point of contention usually relates to whether the specification enables a use of the claimed subject matter. This is close to the question of whether a practical utility is disclosed under 35 U.S.C. § 101. In either case, the Examiner must accept an assertion of utility as true unless there is an objectively reasonable basis to reasonably doubt the asserted utility.[88] If the Examiner advances such reasons, the burden then shifts to the Applicant to provide evidence to substantiate the asserted utility.[89]

In the context of pharmaceutical claims, the Applicant may rely on in vitro data showing pharmaceutical activity in a model that reasonably correlates with the asserted utility in vivo. This does not require the test to be able to predict whether the claimed pharmaceutical product or process or product would actually be useful in treating a disease.[90] Rather, it is generally sufficient if the in vitro testing "establishes a significant probability that in vivo testing for a particular pharmacological activity will be successful.[91]" This can be demonstrated by, for example, providing art showing that the test is accepted as a model system for a particular condition and/or by showing that other similar compounds have the same asserted activity.

---

[87] *In re Wands*, 858 F.2d 731 (Fed. Cir. 1988); M.P.E.P. § 2164.01(a).
[88] *In re Brana*, 51 F.3d 1560, 1566 (Fed. Cir. 1995).
[89] Id.
[90] Id.
[91] *Cross v. Iizuka*, 753 F.2d 1040, (Fed. Cir. 1985).

The case of *In re Brana* is instructive. In that case, the claims related to a class of compounds asserted to have anti-tumor activity, which the Applicant was able to demonstrate in vitro against two specific types of human cancer cells.[92] The Examiner, relying on references that apparently questioned the predictive value of the model systems used by the Applicant, rejected the claims under 35 U.S.C. 112, first paragraph as allegedly lacking enablement because the tests did not establish that such compounds would have antitumor activity in humans.[93] The Federal Circuit reversed, holding that the mere fact that certain references questioned the value of models predict efficacy in humans is irrelevant to whether they are predictive of a pharmaceutical property in vivo. As explained by the Court:

> The references cited by the Board... do not question the usefulness of any compound as an antitumor agent or provide any other evidence to cause one of skill in the art to question the asserted utility of applicants' compounds. Rather, these references merely discuss the therapeutic predictive value of *in vivo* murine tests–relevant only if applicants must prove the ultimate value in humans of their asserted utility. Likewise, we do not find that the nature of applicants' invention alone would cause one of skill in the art to reasonably doubt the asserted usefulness.
>
> The purpose of treating cancer with chemical compounds does not suggest an inherently unbelievable undertaking or involve implausible scientific principles. Modern science has previously identified numerous successful chemotherapeutic agents. In addition, the prior art... discloses structurally similar compounds to those claimed by the applicants which have been proven in vivo to be effective as chemotherapeutic agents against various tumor models.
>
> Taking these facts—the nature of the invention and the PTO's proffered evidence—into consideration we conclude that one skilled in the art would be without basis to reasonably doubt the applicants' asserted utility on its face. The PTO thus has not satisfied its initial burden.
> ***
> Usefulness in patent law, and in particular in the context of pharmaceutical inventions, necessarily includes the expectation of further research and development. The stage at which an invention in this field becomes useful is well before it is ready to be administered to humans. Were we to require Phase II testing in order to prove utility, the associated costs would prevent many companies from obtaining patent protection on promising new inventions, thereby eliminating an incentive to pursue, through research and development, potential cures in many crucial areas such as the treatment of cancer.[94]

In sum, the Examiner may not rest solely on evidence that would question the clinical efficacy of the claimed subject matter, but instead must show that a person of ordinary skill in the art would reasonably doubt the activity of the compound in vivo.

One caveat, however, is that the evidence of enablement must be commensurate in scope with the claim. That is, even though the specification enables some of the subject matter encompassed by the claims, the claim may nonetheless be invalid if there is insufficient guidance to predict which of the remaining subject matter is similarly enabled. Take for example a claim reciting "An isolated peptide, wherein said peptide is at least 75% identical to SEQ ID NO: 1." If the specification discloses that SEQ ID NO: 1 has a particular activity, but does not disclose any

---

[92] *Brana*, 51 F.3d at 1563.

[93] Id. at 1563–64.

[94] Id. at 1566 and 1568.

other species within the scope of the claim, demonstrate the same activity for any other species, there may be a question as to whether the full scope of the claim has been enabled. When faced with such a rejection, it is critical to be able to show something in either the specification or the state of the art that would enable a person of ordinary skill in the art to determine which species have the recited activity. Importantly, though, it is not necessary to show that every species within the claim is operative.[95] For example, where the activity is an enzymatic activity, it may be sufficient to demonstrate either that the specification recites a manner in which the activity could be tested or that such a method was known in the art. Enablement could then be demonstrated by showing that a number of species within the scope of the claim also possess the asserted activity or utility.[96]

From a drafting standpoint, one must draft the application and claims with these issues in mind. As such, it is very important to find not only what the data provided by the inventor show, but why it would be predictive of a particular utility in vivo. Moreover, it is advantageous to include at least a brief discussion in the specification for how this utility could be tested both in vitro and in vivo. It would also be ideal to recite in the specification structural features that are responsible for the asserted utility wherever practical. When presented with a rejection based on a failure to enable a method of using the full scope of the claim, it is important to both attack the Examiner's *prima facie* case and present additional evidence in support of the asserted utility whenever practicable.

## 11.3 Restriction Practice and "Improper Markush" Rejections

In addition to written description, the USPTO has begun urging examiners to narrow the scope of Markush-type claims[97] through both restriction practice and "Improper Markush" rejections.

Pursuant to 35 U.S.C. § 121, "if two or more independent and distinct inventions are claimed in one application, the Director of the USPTO may require the application to be restricted to one of the inventions." The USPTO interprets this as permitting restriction within a single claim.[98] To this end, a procedure has been

---

[95] *In re Angstadt*, 537 F.2d 498, 502–503 (C C P.A. 1976).

[96] See Id. (finding a claim enabled despite encompassing inoperative embodiments where 40 specific embodiments were shown to have the same utility).

[97] A "Markush-type claim" is one that recites at least one group of alternative substituents. M.P.E.P. §§ 803.02 and 2173.05(h). It normally follows a claim structure of "A product/process comprising X, wherein X is selected from the group consisting of...", although other claim structures are acceptable as well. Id. The group introduced by the transitional "selected from the group consisting of" is often referred to as a "Markushgroup". Id.

[98] See 37 C.F.R. § 1.146. This has been strongly criticized as lacking clear support in the statute., see Wegner (2012) Nonetheless, it does not appear that the practice has been challenged in the Courts.

developed for claims containing Markus groups. First, the Examiner must determine whether the claimed subject matter has "unity of invention", which is satisfied as long as all members of the Markush group share a common utility and either a substantial structural feature essential to that utility or[99] If unity of invention is not present, the Examiner may order restriction of the claim to a single invention unless the claimed species are "sufficiently few in number or so closely related that a search and examination of the entire claim can be made without serious burden.[100]" In all other cases, the Examiner may require a provisional election of a single species of invention for examination on the merits.[101] If the elected species is determined to be allowable, examination should be extended to other independent and distinct inventions until either the full scope of the claim is allowed or a narrowed subgenus is allowed.[102]

For pure peptide/protein claims,[103] the USPTO has taken the position that different polypeptides are normally considered independent and distinct chemical species, irrespective of whether they are recited in a single claim.[104] Thus, a pure peptide claim reciting a Markush group of sequences will normally be subject to restriction, absent a linking claim or some evidence showing significantly overlapping structures.[105] As such, the foregoing procedure typically is not applied to pure peptide claims.[106]

To make matters worse, the USPTO recently began urging examiners to reject claims during prosecution if they are found to contain an "improper Markush grouping", arguing that such a rejection is "judicially authorized" if the species lack either (1) a "single structural similarity," or (2) a common use.[107] Additionally, a BPAI panel chaired by Commissioner of Patents Robert Stoll recently ordered an applicant to brief the propriety of a Markush-type limitation, *sua sponte*.[108] It thus is clear that the USPTO intends to treat a proper Markus grouping as a substantive requirement, rather than a requirement as to form. This is bizarre idea, as the courts have acknowledged that the "improper Markush grouping" doctrine lacks any basis in the patent statutes, including 35 U.S.C.

---

[99] M.P.E.P. § 803.02.

[100] Id.

[101] Id.

[102] Id.

[103] By "pure peptide/protein claim", I mean a claim reciting "An isolated peptide/protein comprising an amino acid sequence selected from the group consisting of…" or similar claim structures.

[104] Low and Housel (2012).

[105] Id.

[106] Id.

[107] Supplementary Examination Guidelines for Determining Compliance With 35 U.S.C. 112 and for Treatment of Related Issues in Patent Applications, 76 FR 7162, 7166 (Feb. 9, 2011).

[108] *Ex parte Degrado*, Appeal 2010-005832, App. No. 10/801,951, Order for Further Briefing (BPAI May 9, 2011).

§ 121.[109] As such, subject matter abandoned because of such a rejection may need to be pursued via continuation instead of divisional application, which could render the claims susceptible to obviousness-type double patenting rejections.[110]

All of the foregoing has special consequences for peptide claims. Whereas pure chemical claims can easily be drafted to encompass thousands of species within a single generic structure, the same cannot necessarily be said of peptides. For example, immunogenic peptides that are essentially interchangeable in vaccine compositions by virtue of their specificity for a particular pathogen may nonetheless fail to have any overlapping structure. Thus, if such claims are presented as a Markus group in a peptide, they are susceptible to restriction and/or rejection along sequence lines. This often results in an applicant having to file hundreds or even thousands of applications to cover the full scope of what was invented.

This does not seem to be justified by the patent statute or the case law. As many commentators have agreed, restriction requirements within single claims are not authorized by 35 U.S.C. § 121.[111] Moreover, the case law makes clear that the circumstances under which an "improper Markus grouping" rejection is appropriate should be fairly limited. As the Courts have emphasized, the propriety of a Markush group must be determined by consideration of the claimed subject matter as a whole, not by a rigid comparison of the individual elements.[112] So long as the claim encompasses compounds having "a community of properties justifying their grouping which is not repugnant to principles of scientific classification", the Office must consider the entire scope of the claim.[113] For peptide claims, one can imagine a number of properties that can be encompassed by peptides having completely different sequences such that their classification together is justifiable. For example, the activity of antigenic peptides in an anti-cancer vaccine is determined by, *inter alia*, (1) the ability of the peptide to bind to an appropriate MHC molecule, and (2) the association of the peptide with a protein that is aberrantly expressed or over-expressed in cancerous tissue. A number of scientifically acceptable classifications can be envisioned for such peptides, such as: (1) MHC ligands specific for a given cancer type; (2) MHC ligands derived from the same parent molecule; (3) MHC ligands capable of binding to the same MHC allele, *et cetera*. In none of these cases will the peptide necessarily need to have substantially similar amino acid sequences in order to be classifiable together. Absent an appropriate test case, however, the USPTO does not appear ready to change its practice.

---

[109] See *In re Harnisch*, 631 F.2d 716, 721–22 (CCPA 1980).

[110] See 35 U.S.C. § 121 (reserving divisional applications for inventions for which the Director has required restriction).

[111] Wegner (2012), Tu et al. (2009).

[112] *In re Harnisch*, 631 F.2d 716, 722 (CCPA 1980).

[113] Id.

Nonetheless, some practical steps can be taken in order to obtain a reasonable scope of claims without having to subject to a test case. For example, the USPTO explicitly permits broader Markush groups for composition and/or process (i.e. not a single compound) claims wherein the recited group members "are disclosed in the specification to possess at least one property in common which is mainly responsible for their function in the claimed relationship, and it is clear from their very nature or from the prior art that all of them possess this property.[114]" As such, if the peptide/protein is expected to have a primary commercial viability in a particular context, it may be worthwhile to omit pure peptide claims in favor of claims directed to particular compositions comprising the peptide. For example, where the peptide is primarily for a therapeutic use, the broadest independent claim may be directed to a composition comprising the peptides (or pharmaceutically acceptable salts thereof) and a pharmaceutically acceptable carrier. In such a claim, one would only need to show that the group of peptides has the same or similar function in the composition in order to have an appropriate Markus group. Additionally, the USPTO explicitly permits a reasonable number of independent and distinct species of a generic claim.[115] Oftentimes, examiners will accept a smaller Markus group of distinct sequences, so long as the recited sequences are related in terms of activity and/or source. As such, it may be possible to negotiate with the Examiner to a narrower Markus group. In either case, it is greatly beneficial to recite specific characteristics that unify the various sequences in the specification.

## 11.4 Obviousness

The United States Supreme Court in *KSR v. Teloflex* substantially changed the landscape with respect to obviousness determination. Gone went the mantra that a teaching, suggestion, or motivation in the art is required for obviousness. Instead, a flexible standard based on the guidelines described by *Graham v. Deere* was announced.[116] In particular, the Court explained that the Federal Circuit had consistently erred by rejecting an obvious to try rationale, even when "there are a finite number of identified, predictable solutions, and a person of ordinary skill has good reason to pursue the known options within his or her technical grasp.[117]"

Post-KSR, a number of Federal Circuit cases have addressed the scope of the "obvious to try" rationale. Obvious to try is not appropriate where:

(1) what would have been "obvious to try" would have been to vary all parameters or try each of numerous possible choices until one possibly arrived at a successful result, where

---

[114] M.P.E.P. § 803.02.

[115] 37 C.F.R. § 1.146.

[116] See *KSR Int'l Co. v. Teleflex, Inc.*, 550 U.S. 398, 414 (2007).

[117] Id. at 421.

the prior art gave either no indication of which parameters were critical or no direction as to which of many possible choices is likely to be successful; or

(2) what was "obvious to try" was to explore a new technology or general approach that seemed to be a promising field of experimentation, where the prior art gave only general guidance as to the particular form of the claimed invention or how to achieve it.[118]

For example, the Court has rejected the obvious to try rationale where there is no particular reason to select the closest art from among several equally favorable compounds.[119] In contrast, obvious to try has been affirmed where there is an identifiable reason to modify a known compound using standard and predictable methods.[120] In sum, then, a *prima facie* case of obvious to try requires: (1) a small easily traversable number of starting compounds; and (2) predictability in the results of the modifications to reach the claimed compound from the starting compound. These cases will likely shape the course of claims directed to isolated peptides.

The closest Federal Circuit case to date, *In re Kubin*, stands for the proposition that obviousness of a nucleic acid encoding a known peptide can be established by showing that (1) a person of ordinary skill in the art reasonable likelihood of obtaining the nucleic acid using standard methods; and (2) there is a reason to obtain the isolated nucleic acid.[121] Critically, the Court did not hold that the mere existence of the peptide rendered the nucleic acid obvious; rather it was "because of the peptide's important role in the human immune response" that a person of ordinary skill in the art would be motivated to apply standard methods to identify the nucleic acid encoding the peptide. Thus, it is critical to remember that the Examiner must demonstrate all elements of the claim in the prior art and a motivation or rationale for why a person of ordinary skill in the art would combine the elements.

---

[118] *In re Kubin*, 561 F.3d 1351, Slip Op. at 14–15 (Fed. Cir. 2009).

[119] See 75 FR 53654 (2009) (*citing Takeda Chemical Industries, Ltd. v. Alphapharm Pty., Ltd.,* 492 F.3d 1350 (Fed. Cir. 2007) (not obvious to try when any one of a broad class of compounds could have been selected as the lead compound and the particular modifications necessary to obtain the claimed compound are not predictable); *Ortho-McNeil Pharmaceutical, Inc. v. Mylan Labs, Inc.,* 520 F.3d 1358 (Fed. Cir. 2008) (not obvious when there is no apparent reason why a person of ordinary skill would have chosen the particular starting compound or the particular synthetic pathway that led to the claimed compound and there would be no reason to test for the property possessed by the claimed subject matter); *Sanofi-Synthelabo v. Apotex, Inc.,* 550 F.3d 1075 (Fed. Cir. 2008) (not obvious where the claimed stereoisomer exhibits unexpectedly strong therapeutic advantages over the prior art racemic mixture)).

[120] Id. (citing *In re Kubin*, 561 F.3d 1351 (Fed. Cir. 2009) (claimed polynucleotide is obvious over a known protein that it encodes, there is a reasonable expectation obtaining the claimed polynucleotide using standard biochemical techniques, and a reason to try to isolate the claimed polynucleotide); *Bayer Schering Pharma A.G. v. Barr Labs.,* Inc., 575 F.3d 1341 (Fed. Cir. 2009) ("compound would have been obvious where it was obvious to try to obtain it from a finite and easily traversed number of options that was narrowed down from a larger set of possibilities by the prior art, and the outcome of obtaining the claimed compound was reasonably predictable")).

[121] *In re Kubin*, 561 F.3d 1351, Slip Op. at 14–15 (Fed. Cir. 2009).

The mere recitation of a peptide sequence is not enough to support a rejection directed to such an isolated sequence. Claim 1 in *Ex parte Joo-Eun Bae* recites "an isolated leukemic antigen consisting of a fragment of CD19... or a variant thereof having one, two, or three conservative or non-conservative amino acid substitutions that is capable of stimulating a cytotoxic T lymphocyte reaction against a cell expressing CD19, wherein the fragment is (a) 11–80 amino acids in length or (b) 9–80 amino acids. ..."[122] The claim stood rejected under 35 U.S.C. 102 and 103 over prior art disclosing a 12mer derived from CD19 disclosed solely in the context of a fusion protein.[123] The BPAI held that this did not anticipate the claimed subject matter, as a fusion protein simply is not an "antigen consisting of a fragment of CD19".[124] Moreover, because the sequence was presented solely in the context of a fusion protein, there was no reason to provide it as an isolated antigen.[125] As such, the Examiner failed to state a *prima facie* case of obviousness.

In sum, *KSR* has increased the size of the toolbox available to examiners to reject claims under 35 U.S.C. 103. Nonetheless, the Examiner still bears a heavy burden of persuasion to establish some rationale for why the asserted alteration of the prior art would be predictable and obvious.


# 12 Patenting Peptide-Related Inventions in China[126]

It is well known that the Chinese Patent Law was modeled after the German system, therefore there are a lot of similarities between the two jurisdictions.

There are three sources of legal authority in China with regard to patent. The first and most authoritative is the Patent Law of the People's Republic of China ("the Patent Law"), promulgated by the Chinese People's Congress (the legislature), the latest version took effect on October 1, 2009. Second to that is the Implementation Rules of the PRC Patent Law ("the Patent Rules"), promulgated by China's State Council, and the latest version was issued on January 9, 2010. Finally, China's State Intellectual Property Office ("SIPO"), under the authority of the Patent Rules, establishes a PRC Patent Examination Guidelines for its examination personnel and the public at large. The latest Patent Examination Guidelines were published in January of 2011, and took effect on February 1, 2011. The Examination Guidelines are binding on SIPO personnel, as well as on applicants.

From time to time the Supreme People's Court of China issues opinions or guidelines on the law, which are binding on all courts in China. In the patent law

---

[122] *Ex parte Joo-Eun Bae*, Appeal 2009-013469, Application 10/884,862, Decision on Appeal at 2 (BPAI Apr. 5, 2010).

[123] Id. at 3 and 9–11.

[124] Id.

[125] Id.

[126] By Kening Li, J.D., Ph.D., Pinsent Masons LLP.

area, these opinions can range from how certain provisions of the law should be interpreted (e.g. damage awards) and cases it considered to be "typical", to how the courts should strive to "serve the country" in its judicial activities (e.g. not to adversely impact economic development). These guidelines usually do not have a long-lasting effect, but are very influential at the time they are issued.

## 12.1 Peptide-Related Inventions are Patentable Subject Matter in China

Generally, proteins and peptides themselves are patentable subject matter under the Patent Law, as they are considered to be chemical compositions. The preparation, or making, as well as use, of the peptides are also generally patentable, unless the methods fall into the categories of treatment of humans or animals, or diagnoses of diseases of humans or animals. In China, even in vitro diagnostic methods are excluded from patentability.

Uses of the peptides for the manufacturing of pharmaceutical compositions, or tools, or a kit, and the like, i.e., the so-called Swiss-type claims, are often used to cover methods of treatment or diagnosis.

## 12.2 Novelty and Inventiveness of Peptide-Related Inventions

If the peptides are completely new, then there are no issues of novelty or inventiveness. However, therapeutic peptides, or peptides useful as diagnostic tools, are often part of a protein sequence that has been previously disclosed by another party, especially given the large body of information existed as a result of the Human Genome Project and the associated rush to sequence genomes of various other organisms. These "fragments" of known proteins are patentable even though their sequences were previously known if they have particular functions or activities.

For example, a part of a known protein is found to be a very effective antigenic epitope, and to have application as a therapeutic composition or as a diagnostic tool of a disease. Claims to both the short fragment, and its use for treatment and/or diagnosis purposes are patentable in China. However, a SIPO examiner would likely reject the partial sequence as lack of novelty, citing the published full-length sequence. More frequently, the SIPO examiners would combine the primary sequence that disclose the protein sequence, and a secondary sequence that generally teaches the "usefulness" or a known function of the protein, to reject the partial sequence as lack of inventiveness. Such rejections are improper, however, and with appropriate arguments, they will be withdrawn by the examiner.

In the above example, the inventor has identified a specific portion of the known protein to be the epitope recognized or recognizable by the host's immune system, and as such the short peptide alone is sufficient to elicit a host immune response against the tumor. The specification contains sufficient data supporting the above conclusions.

Because not only the protein sequence was already known, the protein was also known to be associated with the tumor phenotype, the rejection by the examiner of both claims to the peptide and its use in the manufacture of a pharmaceutical composition is not surprising. It is understandable that the examiner may think that the peptide sequence is not novel (its sequence was already known), or alternatively not inventive, because one of ordinary skills in the art would have found it obvious to study the protein, and find a portion thereof as a specific immunogenic antigen, and use it in the manufacture of a treatment composition or as a diagnostic tool.

This argument may seem reasonable but in fact it is not. First of all, the short peptide sequence is novel, because even though the full-length protein molecule was known, there was no teaching in the prior art of the specifically claimed short peptide, which is a completely different chemical entity. Furthermore, the peptide and the associated methods are inventive, because based on the prior art, an ordinarily skilled artisan would not have been able to know if any region of the peptide would actually be immunogenic, and if yes, which specific region. In other words, the short peptide claimed in the application achieves unexpected and superior results.

The claims would be even stronger if the peptides were further modified, e.g,. for improved stability, or formulated to enhance its release profile to achieve better therapeutic effects.

## 12.3 Sufficiency of Disclosure Under Chinese Law

SIPO is known to have a very demanding enablement requirement, especially in the chemical, pharmaceutical and biotechnological arts. The Examination Guidelines emphasize the unpredictability of the "technical effects" in these art areas, and require qualitative or quantitative experimental results sufficiently proving the use or effects of the invention. SIPO examiners are asked to start from the working examples in a patent application for evaluating the scope of claims, and many of them often only look at the working examples with activity data.

When examined closely, however, this "heightened" enablement requirement is not dramatically different from similar requirements in other jurisdictions, notably in Europe. Originally, the intention of SIPO may have been to limit the perceived "land grab" by multinational pharmaceutical companies which were the primary filers of the patent applications in these technical areas, and to leave more breathing room to the Chinese companies hoping to compete. However, as the doctrine develops, and as the Chinese domestic applicants start to file more in these areas, SIPO has more or less adhered to the international norm.

Regardless, the notion that SIPO will require clinical data (in addition to animal testing results), or the specific conditions for administering a pharmacological compound, is incorrect. Plenty of pharmaceutical patents have been issued in China with no such clinical data, or even animal test data on the claimed compositions.

More often than not, individual examiners assert this "heightened requirement" due to his or her lack of familiarity with the particular technical area. A clear explanation of the background art, and what an ordinarily skilled artisan would consider easily achievable, coupled with artful drafting of the specification, can overcome these kind of rejections.

It goes without saying that when drafting such a patent application, it is highly desired to include as much data as possible. Also, it is useful to include clear, explanatory, albeit qualitative statements about the claimed invention. Prophetic data are helpful, if indeed supported by the invention. If the claimed invention involves alternative components, statements that all claimed components have the claimed activity are highly helpful. If the claimed invention involves a data range, experimental data that show that the lowest and highest data points, or at least qualitative statements to the same effect, will be very useful in the future to support the enablement of the range.

## 12.4 Submission of Post-Filing Data

Among patent prosecution professionals, both Chinese and international, there is a further erroneous belief that SIPO does not accept any post filing data.

Although it is true that SIPO, as any other patent authority in the world, does not allow any addition of new matter to the original disclosure, post-filing experimental data, if presented appropriately, can be and often are used to over-come both lack-of-inventiveness rejections and lack-of-sufficient-disclosure rejections.

For one thing, the Examination Guidelines explicitly allow "commercial success" as part of the evidence to support inventiveness of an invention. As commercial success, by definition, must be post filing in a first-to-file country such as China, it is clearly incorrect to say that post-filing data cannot be used. Comparative studies between the claimed composition and a close, but inferior prior art composition, to show superior and unexpected results are routinely used, and accepted by SIPO as evidence of inventiveness of the invention.

Using post-filing data to support enablement is less straightforward, especially in unpredictable arts such as chemical compounds and pharmaceutical uses of the compounds. There should not be too much difficulty to use data generated post filing to support or supplement statements made in the patent specification about alternative embodiments. For example, if the invention is stated to work with components A, B, C, and D, actual data are provided with regard to A, and a skilled artisan would recognize that given the shared characteristics of A, B, C, and D, the other, not actually tested alternatives, should also work in the claimed invention. If the examiner should raise any rejection as to the lack of disclosure with regard to B, C, and D, the applicant should easily overcome the rejection by submitting data actually proving the B, C, and D works.

In less predictable art, such as the use of various peptide vaccines for the treatment of a disease, the specification should provide a credible rationale or scientific theories to support the extrapolation of the data from one peptide to another, and make definite statements about the alternatives that are not supported by actual data. These kinds of explanations and statements would give the applicant the basis to submit and use post-filing data to prove to the examiner that the statements made in specification at the time of the filing were correct.

## 12.5 Unity of Invention

A common and often difficult issue is whether unity of invention exists between a polypeptide and a polynucleotide that encodes the polypeptide. For an examiner the U.S. PTO often insists that unity does not exist between the two, even though the PCT Rules explicitly state otherwise.

The Chinese SIPO adheres to the PCT rules, and accepts the unity between the apolynucleotide and the polypeptide encoded by the polypeptide, or a polypeptide and any polynucleotide that encodes the polypeptide, with the degeneracy of the genetic code taken into consideration. Similarly, a cell that comprises the polypeptide, or the polynucleotide, is also considered to have unity. Also, an antigen and any antibody generated using the antigen would be considered to have unity.

A related issue is whether it exists among polypeptides or polynucleotides that share certain sequence homology. Here, SIPO applies the general principle that the inventions should share a common technical feature, which defines the inventions over the prior art. Thus, if the sequences share a common structural "core", as well as a common function, they would be considered to have unity.

The sequence homology may be defined by either percent identity, or in languages such as "one more conservative substitutions." In either case, the specification must contain sufficient data to support the assertion that the homologous or substituted sequences do retain the function. For example, if the claim recites that "or a polypeptide that is at least 95% identical to SEQ ID NO: 1, and has function X", then the specification should have experimental data or other proof that at least one molecule that is 95% homologous does have the same function. Similarly, if a short peptide is recited to have SEQ ID NO: 1 with one or two conservative substitutions, data should be presented to support this statement. Otherwise, the claims will be rejected for both lack of unity of invention, and lack of sufficient disclosure. In other words, claims to sequences simply based on homology, without data to support their function are generally rejected.

SIPO's Examination Guidelines provide the following examples:

1. A purified antibody to protein A or protein B.
   The specification provides that protein A and protein B have completely

different structures, and the antibodies to protein A and protein B do not have any common function. The claim lacks unity.

2. A purified growth factor peptide selected from the group consisting of SEQ ID NOs: 1–10.

   The specification provides that SEQ ID NOs.1–10 have a common function to promote the growth of a cell, but do not have any conserved activity/structural domain. The claim lacks unity.

3. A peptide inhibitor of protein C selected from the group consisting of ARDNCEQGHIL and ARDNCEQKMIL.

   The specification provides that the RDNC domain was not previously identified, and is necessary for a higher inhibitory activity. Since the two peptides have a common activity/structural domain, and this domain is necessary for stronger inhibition and distinguishes the claimed peptides from the prior art, the claim meets the unity requirement.

## 12.6 Experimental Use Exemption to Patent Infringement

An area of concern to international patent applicant in China is the strength of patent protection. There are many obvious issues beyond the scope of this note. One area, however, is highly relevant to applicants in the pharmaceutical arts, that is, the so-called "Bolar Exemption" enacted in China's Third Amended Patent Law, which took effect on October 1, 2009.

In its early days, China's patent law defined "experimental use" very broadly, so much so that any use for scientific research purposes were exempted. This controversial position gradually gave to the current, narrow definition of "experimental use", which encompasses only testing of the patented technology, and use of the patented technology for pure scientific research purposes without any commercial objectives. Thus, the use of a patented technology to develop another technology is not exempted.

In order to obtain market approval, data must be generated and submitted to the regulatory authority (i.e. China's State Food and Drug Administration, or SFDA) that the pharmaceutical composition or diagnostic product is safe and effective. Generation of such data often takes a long time, and it would be unfair to the applicant to wait until the underlying patent(s) to have expired before a generics applicant can conduct such tests. Thus, like many other countries, China exempts activities "solely for the purpose of obtaining regulatory approval" from patent infringement. This is the so-called "Bolar Exemption", contained in the Hatch-Waxman Act in the US.

China's Bolarexemption, however, is not balanced by either the patent-term extension or the patent linkage mechanism. The patent-term extension compensates the branded pharmaceutical companies for the loss of its patent term due to the lengthy approval process it had to endure, while patent linkage provides a mechanism for the patentee to be timely notified of any intent by a generic company to infringe its patents, and to initiate court proceedings to stop any potential infringement. Both patent-term extension and patent linkage are universally considered to be necessary in balancing the interests of the patentee and those of the public and the generics companies.

Because the Chinese version of the Bolar Exemption is one-sided, favoring the generics companies, it is one of the most criticized provisions of the current Chinese patent law.

# 13 Conclusion

As China's economy develops, and as an increasing number of Chinese companies are seeking patent protection of their own innovations, the Chinese patent system has increasingly become more in line with the generally accepted international practices. Experiences suggest that the difficulties a Western patent applicant has before the SIPO are more frequently due to the lack of adequate advocacy skills of the Chinese patent agents, who, compared to their Western counterparts, are more used to accepting the assertions made by the patent examiners. Another factor may be due to inaccurate or even incorrect translation. In complex art areas such as peptide-related inventions, the applicant should provide a more detailed explanation of the background of the invention and when appropriate, perhaps the mechanism of the invention, so as to minimize lack of enablement-related rejections by the examiner.

**Acknowledgments** This chapter comprises a contribution related to aspects of US law written by Bryan W. Jones, Baker, Donelson, Bearman, Caldwell & Berkowitz, PC, Washington, and a contribution related to aspects of Chinese law written by Kening Li, Pinsent Masons LLP, Shanghai.

# References

Ambrogelly A, Palioura S, Söll D (2007) Natural expansion of the genetic code. Nat Chem Biol 3(1):29–35
Coligan et al (eds) (2000) Current protocols in protein science. Wiley, New York
Reinhardt C, Zdrojowy R, Szczylik C, Ciuleanu TE, Brugger W, Oberneder R, Kirner A, Walter S, Singh H, Stenzl A (2010) Results of a randomized Phase 2 study investigating multi-peptide vaccination with IMA901 in advanced renal cell carcinoma (RCC). J Clin Oncol, ASCO Ann Meeting Proc Part I. 28: 4529
Cotropia et al (2010) Do applicant patent citations matter? Implications for the presumption of validity, second annual research roundtable on the empirical studies of patent litigation (Nov. 2010), available at http://www.law.northwestern.edu/searlecenter/papers/Cotropia_patent_citations.pdf
Dengjel J, Nastke MD, Gouttefangeas C, Gitsioudis G, Schoor O, Altenberend F, Müller M, Krämer B, Missiou A, Sauter M, Hennenlotter J, Wernet D, Stenzl A, Rammensee HG,

Klingel K, Stevanović S (2006) Unexpected abundance of HLA class II presented peptides in primary renal cell carcinomas. Clin Cancer Res 12:4163–4170

Gnjatic S, Atanackovic D, Jäger E, Matsuo M, Selvakumar A, Altorki NK, Maki RG, Dupont B, Ritter G, Chen YT, Knuth A, Old LJ (2003) Survey of naturally occurring CD4+T-cell responses against NY-ESO-1 in cancer patients: correlation with antibody. Proc Natl Acad Sci U S A 100(15):8862–8867

Kennedy RC, Shearer MH, Watts AM, Bright RK (2003) CD4 + T lymphocytes play a critical role in antibody production and tumor immunity against simian virus 40 large tumor antigen. Cancer Res 63:1040–1045

Kobayashi H, Omiya R, Ruiz M, Huarte E, Sarobe P, Lasarte JJ, Herraiz M, Sangro B, Prieto J, Borras-Cuesta F, Celis E (2002) Identification of an antigenic epitope for helper T lymphocytes from carcinoembryonic antigen. Clin Cancer Res 8:3219–3225

Leber TM (2009) Unity of chemical and biotechnological Markush Claims under the PCT and EPC—consistency of the PCT and EPC guidelines with the law (IIC 2009, 206)

Low C, Housel J (2012) Restriction Practice Guide, available at www.uspto.gov/web/patents/biochempharm/documents/low.pps (last Accessed Feb. 7, 2012)

Lundblad R (2005) Chemical Reagents for Protein Modification, 3rd edn. CRC Press

Meziere C, Viguier M, Dumortier H, Lo-Man R, Leclerc C, Guillet JG, Briand JP, Mulle S (1997) J Immunol 159: 3230–3237

Qin Z, Blankenstein T (2000) CD4 + T-cell–mediated tumor rejection involves inhibition of angiogenesis that is dependent on IFN gamma receptor expression by nonhematopoietic cells. Immunity 12:677–686

Qin Z, Schwartzkopff J, Pradera F, Kammertoens T, Seliger B, Pircher H, Blankenstein T (2003) A critical requirement of interferon gamma-mediated angiostasis for tumor rejection by CD8 + T cells. J Cancer Res 63(14):4095–4100

Rammensee HG, Falk K, Rotzschke O (1993) Peptides naturally presented by MHC class I molecules. Annu Rev Immunol 11:213–244

Rammensee HG, Bachmann J, Stevanovic S (1997a) MHC ligands and peptide motifs. Landes Bioscience, USA

Rammensee HG, Bachmann J, Stevanovic S (1997b) Ligands and Peptide Motifs. Springer, Heidelberg

Rammensee HG, Bachmann J, Emmerich NP, Bachor OA, Stevanovic S (1999) SYFPEITHI: database for MHC ligands and peptide motifs. Immunogenetics 50:213–219

Singh H, Hilf N, Mendrzyk R, Maurer D, Weinschenk T, Kirner A, Frisch J, Stenzl A, Reinhardt C, Walter S (2010) Correlation of immune responses with survival in a randomized phase 2 study investigating multi-peptide vaccination with IMA901 plus/minus low-dose cyclophosphamide in advanced renal cell carcinoma (RCC). Journal of Clinical Oncology, 2010 ASCO Annual Meeting Proceedings Part I. 28: 2587

Singh-Jasuja H, Walter S, Weinschenk T, Mayer A, Dietrich PY, Staehler M, Stenzl A, Stevanovic S, Rammensee V, Frisch J (2007) Correlation of T-cell response, clinical activity and regulatory T-cell levels in renal cell carcinoma patients treated with IMA901, a novel multi-peptide vaccine; ASCO Meeting 2007 Poster # 3017

Staehler M, Stenzl A, Dietrich PY, Eisen T, Haferkamp A, Beck J, Mayer A, Walter S, Singh H, Frisch J, Stief CG (2007) An open label study to evaluate the safety and immunogenicity of the peptide based cancer vaccine IMA901, ASCO meeting 2007; Poster # 3017

Stewart M, Kent L, Smith A, Bassinder E (2011) The special inventive step standard for antibodies. EPI Inf 2:72

Tu S, et al (2009) Squeezing more patent protection from a smaller budget without compromising quality, landslide 2(2)

Viatte S, Alves PM, Romero P (2006) Immunol Cell Biol 84(3):318–330

Wegner (2012) The Eagle Right to Generic Protection, available at http://www.grayonclaims.com/storage/EagleRighttoGenericClaims.pdf (last Accessed 16 Jan, 2012)

# Patent Landscape in Molecular Diagnostics

Johanna Driehaus

**Abstract** According to latest figures the molecular diagnostic market in the US alone is worth about $2.9 billion with a predicted annual growth of 15% until 2015 resulting in a volume of $6.2 billion. Specifically, potential is seen in genomic diagnostics due to the now readily available next generation sequencing techniques for sequencing of individual cancer genomes. Thus, this chapter gives an overview of the patent landscape in molecular diagnostics, and discusses issues of patentability with respect to the different technologies and compounds used therein.

**Keywords** Pyrosequencing · DNA methylation · Forensics · Disease detection · Personalized medicine · Biological compounds · Drafting recommendations

## 1 Introduction

According to latest figures the molecular diagnostic market in the US alone is worth about $2.9 billion with a predicted annual growth of 15% until 2015 resulting in a volume of $6.2 billion.[1] Specifically, potential is seen in genomic diagnostics due to the now readily available next generation sequencing techniques

---

[1] *Source* The Future of Molecular Diagnostics: Innovative technologies driving market opportunities in personalized medicine, 23rd June 2010. Report published by Business Insights Ltd.

---

J. Driehaus (✉)
Viering, Jentschura & Partner, Kennedydamm 55, 40476 Duesseldorf, Germany
e-mail: jdriehaus@vjp.de

U. Storz et al., *Intellectual Property Issues*, SpringerBriefs in Biotech Patents, DOI: 10.1007/978-3-642-29526-3_3, © The Author(s) 2012

73

for sequencing of individual cancer genomes. Thus, this news alone justifies a closer look at the relevant patent issues accompanying these developments.

Typically, when molecular diagnostics is discussed in the context of patents, it mainly refers to disease gene patents. However, herein, molecular diagnostic patents not only cover disease gene patents but also those patents claiming tests and methods to identify and possibly treat a disease or the predisposition for a disease on DNA, RNA, or even the protein level of an organism.

## 2 Legal Background

Most countries have excluded diagnostic methods from the scope of their patent systems. The Agreement on Trade Related Aspects of Intellectual Property Rights (TRIPS) under Paragraph 3 of Article 27 allows members to exclude diagnostic, therapeutic, and surgical methods for the treatment of humans or animals from the scope of patentable subject matter.

In case of the US, the paradigm of the US Patent and Trademark Office (USPTO) to determine the patentability of inventions involving nanotechnology and biotechnology continues to evolve due to the Supreme Court decision in the case of *Bilski v. Kappos*, 2010 WL 2555192.

Throughout the litigation history of this case, the focus was on the 'machine or transformation' requirements articulated by the USPTO and further solidified by the Federal Circuit: "A claimed process is surely patent-eligible under § 101 if: (1) it is tied to a particular machine or apparatus, or (2) it transforms a particular article into a different state or thing." Essentially, the legal issue generating all the consternation and strife was: "What test or set of criteria governs the determination by USPTO or courts as to whether a process is patentable?" The Federal Circuit opined that a process is not patentable subject matter unless it exactly conforms to the 'machine or transformation' test. However, the Supreme Court relegated the 'machine or transformation' test to the role of a helpful clue in determining whether or not a process is patentable subject matter. The bottom line, as it stands now, is that patentable subject matter is still quite broad, and the analysis to determine it includes the 'machine or transformation' test as one criterion with regard to processes, but it is not the sole deciding factor. If the Supreme Court had required strict adherence to the machine or transformation test, it would have had seismic implications for patents already issued, and certainly would have caused trouble for pending applications relating to biotechnology and nanotechnology. The trouble arises from the precarious position that biotech and nanotech patents are not theoretically involved with a machine and nor do they transform matter. Now, there is a least the possibility for these types of inventions to move forward in the process of patentability.

In the case of the EPC former Article 52(4) now Article 53(c) EPC excludes diagnostic methods practiced on the human or animal body from patentability. Article 53(c) was included in the EPC to protect medical and veterinary practitioners from infringing patents relating to such diagnostic methods in the course of their work.

In detail, the Enlarged Board of Appeal (EBA) considered a diagnostic method to be a multi-step process consisting of: data collection, comparison of data with standard values, finding of significant deviation, and attribution of the deviation to a particular clinical picture. Accordingly, all these steps must be present for such a method to be considered a method of diagnosis and therefore be excluded.

The EBA also looked at the interpretation of "practiced on the human or animal body". It considered that a technical method step satisfied the criterion of being "practised on the human or animal body" if the method implied any interaction necessitating the presence of the human or animal body. Therefore, any kind of interaction with the human or animal body, whether it is invasive or noninvasive, is considered to be "practiced on the human or animal body". However, methods which do not require interaction with the human or animal body are not methods of diagnosis. For example, method steps carried out in vitro using ex vivo samples.

The Boards of Appeal seem to follow the guidelines set by the G01/04 resulting in a narrow interpretation of Article 53(c) EPC. For example, the decision T 1255/06 concerned a radiation detector for tympanic temperature measurement. Claim 16 was directed at

A method of determining ear temperature, comprising the steps of (...); inserting the extension (18) into an ear; and detecting the radiation, (...)and converting the peak radiation sensed to a peak ear temperature.

The board argued that while acquisition of the temperature data leads to the detection of a deviation from the normal values, it does not allow per se the attribution of the detected deviation to a particular clinical picture. Therefore, the claim does not define the features relating to the diagnosis for curative purposes stricto sensu. Hence, the claim did not fall under the exclusion of Article 53(c) EPC.

Apart from the legislative texts of the patent laws, molecular diagnostics are also subject to various other regulations, e.g., approval by the Food and Drug Administration (FDA). The FDA is an agency of the United States Department of Health and Human Services, responsible for protecting and promoting public health through regulation and supervision. Especially relevant to the field of molecular diagnostics are the recent changes in the FDA's approach toward genetic tests. For many years there has been something of a set of double standards in the way in which in vitro diagnostic tests have been regulated in the US. On the one hand in vitro diagnostic products—kits that are on sale to laboratories—have, in general, been very strictly regulated by the FDA. Extensive testing and validation has been required before approval for sale could be obtained. In contrast, diagnostic assays that laboratories develop themselves—the so-called "home brew" assays—have effectively by-passed these strict FDA regulation and fallen under an alternative very light regulation.[2] Under new stricter regulations which are currently set up, genetic tests sold directly to consumers require regulatory approval as medical devices before they can be marketed.[3]

# 3 Key Technologies

In the following sections several tables with key IP rights in the US and their European counterparts are presented. In these tables, the column "end of 20 year term" shows the year in which 20 years since the filing date of the respective US patent application have passed. If the date differs for the European patent application, the year is shown below the patent application number. This term is given, since at least in Europe patent protection (with the exception of supplementary protection certificates) runs out 20 years after the filing date at the latest. Due to differences in the law between the US and Europe and associated liability risk it is not possible to show the actual end of the patent term. However, the given date can be used as a rough guideline to determine whether the patents that could have been derived from the patent applications might still be in force. Moreover, for some patent applications other details are given, for instance if a European patent has been revoked.

## 3.1 Polymerase Chain Reaction

The Polymerase Chain Reaction (PCR), invented in 1983 by Dr. Kary B. Mullis while working at the Cetus Corporation, was a ground-breaking technique enabling amplification of a single or few copies of a piece of DNA across several orders of magnitude.

The first patent covering the application of the PCR-technique was filed in 1985 by Cetus (Ser. No. 716975) and in the same year the first article describing PCR amplification of the human beta globin genes was published in Science.[4] However, the first PCR method had the drawback that the used DNA polymerase—the Klenow fragment—was not thermostable. Thus, after each cycle new enzyme had to be added to the reaction.

---

[2] Little (2006).

[3] *Source* The New York Times, June 11, 2010.

[4] Saiki et al. (1985).

**Table 1** Key IP rights protecting PCR

| Company | Technology | End of 20 year term | Key IP right US | Key IP right EP |
|---|---|---|---|---|
| Roche molecular systems | PCR setup | 2005 | US4683202 | EP0201184 |
| Roche molecular systems | PCR primers | 2005 | US4683202 | EP0505012 |
| Roche molecular systems | PCR machine | 2010 | US5656493 | EP0236069 revoked |
| Roche molecular systems | Taq enzyme | 2007 | US4965188 | EP0258017 |

Therefore, shortly after the publication in Science, Mullis group began to use the thermostable DNA polymerase "Taq" instead of the Klenow fragment, thereby eliminating the need of having to add new enzyme to the PCR reaction during the thermocycling process. Since then a single closed tube in a relatively simple machine can be used to carry out the entire process. Hence, the use of Taq polymerase, inter alia protected by EP0258017, was the key idea that made PCR applicable to a large variety of molecular biology problems concerning DNA analysis.

After establishing the PCR method, Cetus in 1985 formed a joint venture with the Perkin-Elmer Corporation in Norwalk, Connecticut, and introduced the DNA Thermal Cycler.

Two years later, in 1987 Cetus Corporation received the patent rights for the PCR technique from the United States Patent Office (USPTO) and by 1988 Cetus was receiving numerous inquiries about licensing to perform PCR for commercial diagnostic purposes.

On January 15th 1989, Cetus announced an agreement to collaborate with Hoffman-LaRoche on the development and commercialization of in vitro human diagnostic products and services based on PCR technology.

In 1991 Hoffmann-La Roche acquired the worldwide IP rights for PCR for $300 million, while Cetus merged with Chiron Corporation. Roche made it easier to acquire a license for PCR. It established different categories of licenses related to PCR, depending on the application and the users. These categories included research applications, such as the Human Genome Project, the discovery of new genes, and studies of gene expression, diagnostic applications, such as the detection of disease-linked mutations, the production of large quantities of DNA, and the most extensive PCR licensing program, human diagnostic testing services. Licenses in the last-named category were very broad, there were no upfront fees or annual minimum royalties, and the licensees had options to obtain reagents outside Roche. Overall Roche earned approximately $350 million with the PCR royalties per year.

Due to this licensing scheme the discussion about access to PCR technology centered on the costs of Taq polymerase, used in the amplification, rather than on the distribution of intellectual property rights.

However, in 1999 the US4889818 Taq patent was ruled unenforceable in the US due to misleading information and false claims. The terms of all key IP rights listed in Table 1 should have expired in the US and Europe by the editorial deadline of this volume.

## 3.2 Reverse Transcriptase PCR

In 1970, the scientists Howard Temin and David Baltimore both independently discovered the enzyme responsible for reverse transcription, named reverse transcriptase ("RT"). It is a naturally occurring enzyme produced by retroviruses, such as the Moloney-Murine Leukemia Virus ("MMLV").

During cDNA synthesis the reverse transcriptase degrades the mRNA strand of the mRNA/cDNA hybrid molecule, a process termed RNase H activity, so that the first strand cDNA nucleotides are free to form a second strand and complete the DNA replication. However, if RNase H activity destroys the mRNA template, as it happens with naturally occurring RT, then it cannot serve as a template for additional cDNA. Thus, an RT with inhibited RNase H behavior is useful for more efficient reactions.

Therefore, mutant RT with DNA polymerase, but no RNase H activity ("RNase H minus") was developed and the corresponding patent application US07/143,396 was filed in 1988 by Invitrogen's Life Technologies division. From the parent application of 1988 the patents US6063608, US5668005 and US5244797 were derived, which cover significantly improved reverse transcriptases that increase the length, yield, and quality of cDNA produced from mRNA in a reverse transcription reaction.

However, beginning in the early 1980s two scientists at Columbia University, Dr. Stephen P. Goff and his post-doctoral researcher, Dr. Naoko Tanese, studied the effects of random mutations in the MMLV gene for RT. Two mutant genes created in 1984 were H7 and H8, each encoding enzymes that later proved to lack RNase H activity. In late 1984, Tanese even tested the mutant RT for RNase H activity, but these tests using 1984 assay technology yielded inconclusive results. When Goff and Tanese completed a new in situ assay in March 1987, they rapidly determined which parts of the MMLV RT gene affected which enzyme properties. By March 7, 1987, they had established that H7 and H8 encoded mutant RT with DNA polymerase activity but no RNase H activity. Goff and Tanese started publishing their work in March 1987, so before the priority date of Invitrogen's US07/143,396. On January 29, 1988, Goff filed a patent application pertaining to this research. Consequently, the USPTO declared an interference between Goff's application and Invitrogen's US5668005 patent in 1993 (October 18, 1993, notice of interference from USPTO). In contrast to most other countries, which award patent rights based on the first person to file a patent application for an invention, the US awards patent rights based on the first person to invent, even if that person filed his or her patent application after another person. A first-to-invent dispute can be resolved in an interference proceeding at the USPTO, wherein a determination is made as to which party has priority in the invention, i.e., who was the first to invent. However, Goff's assignee, Columbia University, defaulted and the USPTO ruled in Invitrogen's favor. As a result, the USPTO never reviewed Goff's research records to determine priority of invention between Goff and Invitrogen.

Invitrogen offers more than 100 different products based on the US6063608, US5668005, and US5244797 patents. Via purchase of any of the products covered by the patents Invitrogen provides its customer with a limited license under its patents to use them for research purposes. Such Invitrogen products include SuperScript[TM] RT and ThermoScript[TM] RT, kits containing these reverse transcriptases and cDNA libraries made with them. However, Clontech's PowerScript products and their customers' use of them are not licensed under these patents.[5] Therefore, in 1996 Invitrogen sued Clontech for infringement of the patents US5244797, US5668005 and US6063608.

In the subsequent proceedings the US District Court for the District of Maryland invalidated 221 claims, in the three related Invitrogen patents as anticipated by prior art, i.e., by the research of Goff and Tanese. Invitrogen appealed and the US Court of Appeals for the Federal Circuit found that the PowerScript products sold by Clontech infringed Invitrogen's US6063608. In the decision, the Court of Appeals remanded the case to the district court for further proceedings. Finally, in 2007 Invitrogen and Clontech jointly announced a confidential settlement of the patent litigation. As part of the settlement, Clontech has agreed that Invitrogen's US5244797, US5668005, and US6063608 are valid and enforceable. As a result Clontech discontinued sales of its RNase H minus RT products, including its PowerScript products, for the life of the patents. The parties did not disclose other details of their agreement.[6]

Moreover, RT-PCR, which uses a single thermostable polymerase - rTth - was developed in 1991. This achievement greatly simplified the RT-PCR procedure, which could now take place in a single tube.

Also in 1991 the first thermostable RT-PCR research kit was launched and Hoffmann-La Roche Inc. subsequently acquired worldwide rights and patents, e.g. US5322770, to the technique.[7] Table 2 shows key IP rights protecting reverse transcriptase PCR.

---

[5] *Source* Press release of December 01, 2005.

[6] Albainy-Jenei (2005).

[7] Rajan (2009) **and** Roche PCR Timeline at http://molecular.roche.com/roche_pcr/pcr_timeline. html.

**Table 2** Key IP rights protecting reverse transcriptase PCR

| Company | Technology | End of 20 year term | Key IP right US | Key IP right EP |
|---|---|---|---|---|
| Invitrogen | Reverse transcriptase lacking | 2017 | US6063608 | None |
| | | 2012 | US5668005 | None |
| | RNase H activity (all derived from No. parent application 07/143, 396) | 2013 | US5244797 | None |
| Hoffmann-La Roche Inc | Thermostable reverse transcriptases | 2011 | US5322770 | EP0550687 |

## 3.3 Real-Time PCR

One of the further developments of the PCR technology is quantitative real-time PCR. The procedure follows the general principle of polymerase chain reaction. Its key feature is that the amplified DNA is detected as the reaction progresses in *real time*, a new approach compared to the standard PCR, where the product of the reaction is detected after the reaction is completed. Two common methods for detection of products in real-time PCR are: (1) non-specific fluorescent dyes that intercalate with any double-stranded DNA (claimed for example in the US6569627) and (2) sequence-specific DNA probes consisting of oligonucleotides that are labeled with a fluorescent reporter, which permits detection only after hybridization of the probe with its complementary DNA target (claimed for example in the US5210015).

Two companies have acquired the core IP rights for the technologies. While Roche received the patent US6171785 and its European counterpart EP0512334 for the key reagents required in the reaction, Applera holds the patents US6814934 and EP0872562 covering the necessary laboratory equipment.[8] The EP0872562 was subject to several litigations but at present is upheld in amended form.

In a further development of the technique Idaho Technology in collaboration with the University of Utah in 1996 launched the LightCycler® Instrument, a rapid thermal-cycler with a built-in fluorescence detection system for real-time gene quantification. The LightCycler Instruments allow users to complete typical DNA amplification reactions and analyze the results, in less than 30 min. In 1997, Idaho Technology sublicensed the technology to Roche Diagnostic and entered into a multi-year research agreement to develop innovative new products. Table 3 shows key IP rights protecting real-time PCR.

---

[8] Van Guilder et al. (2008).

**Table 3** Key IP rights protecting real-time PCR

| Company | Technology | End of 20 year term | Key IP right US | Key IP right EP |
|---|---|---|---|---|
| Roche diagnostic | Real-time PCR reagents | 2020 | US6171785 | EP0512334 |
| Originally Applera now applied biosystems | Real-time cycler | 2017 | US6814934 | EP0872562 |
| University of Utah | Rapid thermal cycling | 2018 | US6787338 | EP0906449 |
| Roche | Taqman | 2010 | US5210015 | EP0543942 |
| University of Utah | Use of dyes SYBR Green I in qPCR | 2021 | US6569627 | None |
| University of Utah (licensed first to idaho technology then to roche diagnostics) | Monitoring nucleic acids with probes or dyes during or after amplification i.e., basic for SYBR Green and FRET technology | 2017 | US6174670 | EP1179600 |

## 3.4 Sequencing

The term DNA sequencing refers to sequencing methods for determining the order of the nucleotide bases—adenine, guanine, cytosine, and thymine—in a molecule of DNA. Prior to the development of rapid DNA sequencing methods in the 1970s by Frederick Sanger at the University of Cambridge and Walter Gilbert/Allan Maxam at Harvard, a number of laborious methods were used.

Until automated sequencing instruments were widely available, only a few laboratories had access to this technology. The prototypes for these instruments were developed by Leroy Hood and colleagues at Caltech in the years 1980–1986. During this time the team of scientists increased the sensitivity of protein sequencing instruments by a factor of about 100. They patented their findings 1990 and were granted the US5171534 in 1992. This work was only possible with support from the private sector, and companies were apparently very reluctant to invest in developing the sequencing instrumentation. Hood approached 19 companies, all of which declined to support the development of the sequencers. Eventually, he obtained funding from Applied Biosystems (ABI) and ABI insisted on, and received, an exclusive license.[9]

Patent issues concerning DNA sequences became widely known in the course of the Human Genome project. This was especially due to Craig Venter, who was a scientist at the NIH during the early 1990s when the project was initiated. His firm Celera used a technique called whole genome shotgun sequencing, related to the Sanger sequencing method, differing from the otherwise used map

---

[9] Intellectual Property Rights and Research Tool in Molecular Biology, Summary of a workshop held by the National Academy of Science, February (1996).

**Table 4** Key IP right protecting sequencing techniques

| Company | Technology | End of 20 year term | Key IP right US | Key IP right EP |
|---|---|---|---|---|
| California Institute of Technology licensed to Applied Biosystems | Automated sequencer | 2010 | US5171534 | None |

and then shotgun technique.[10] Instead of sharing their data with the public funded teams working on the project Venter and Celera announced that they would seek "intellectual property protection" on "fully-characterized important structures" amounting to 100–300 targets. The firm eventually filed preliminary ("place-holder") patent applications on 6,500 whole or partial genes thereby receiving substantial public attention.[11] However, in 2005 Celera placed its formerly proprietary genome sequence data into the public domain, thereby ending the first grand business experiment in the era of commercial genome sequencing. This was also due to the fact that during 2004, the US, European and Japanese patent offices became significantly more restrictive in their requirements for patenting DNA stating that a given sequence can only be protected, if a specific function is disclosed. Hence, Celera ultimately failed in patenting the sequenced DNAs.[12] Table 4 shows a key IP right protecting sequencing techniques.

## 3.5 Pyrosequencing

Pyrosequencing is a method of DNA sequencing based on the "sequencing by synthesis" principle. Essentially, the method allows sequencing of a single strand of DNA by synthesizing the complementary strand along it, one base pair at a time, and detecting, which base was actually added at each step. Light is produced only when the nucleotide added—each nucleotide is added in turn—complements the unpaired base of the template. The sequence of solutions, which produce chemiluminescent signals, allows the determination of the sequence of the template.

The company Pyrosequencing AB in Uppsala, Sweden commercialized machinery and reagents for sequencing short stretches of DNA using the pyrosequencing technique. Pyrosequencing AB was renamed as Biotage in 2003, which was acquired by Qiagen in 2008. Pyrosequencing technology was further

---

[10] Goodmann (1998).

[11] BBC News Wednesday, 27 October, 1999.

[12] Kling (2005).

**Table 5** Key IP rights protecting pyrosequencing techniques

| Company | Technology | End of 20 year term | Key IP right US | Key IP right EP |
|---------|------------|---------------------|-----------------|-----------------|
| Pyrosequencing AB | Methods for sequencing DNA, where the incorporation of bases is coupled to the release of pyrophosphate (PPi) and its detection via a light signal | 2017 | US6210891 | EP0932700 |
| Pyrosequencing AB | Improved DNA sequencing method involving degradation of the unincorporated nucleotides through the addition of a nucleotide-degrading enzyme | 2017 | US6258568 | EP0946752 |

licensed to 454 Life Sciences. 454 developed an array-based pyrosequencing technology, which has emerged as a platform for large-scale DNA sequencing. Most notable are the applications for genome sequencing and metagenomics. *GS FLX*, the latest pyrosequencing platform by 454 Life Sciences (now owned by Roche Diagnostics), can generate 400 million nucleotide data in a 10 h run with a single machine. Each run costs about $5,000–7,000, pushing de novo sequencing of mammalian genomes into the million dollar range. Table 5 shows key IP rights protecting Pyrosequencing techniques.

## 3.6 Nucleic Acid Extraction

Obtaining high quality, intact nucleic acid is the first and often the most critical step in performing many fundamental molecular biology experiments, including genotyping, PCR, and cDNA library construction.

### 3.6.1 General Methods

The separation of RNA from DNA is often carried out using an acid phenol/guanidine mixture commonly known as Trizol, DNAzol or QIAzol, etc. When employing these reagents a biological sample is homogenized in an aqueous solution of phenol and guanidine isothiocyanate, thereafter the homogenate is mixed with chloroform. Following centrifugation, the homogenate separates into an organic phase, an interphase, and an aqueous phase. Proteins are sequestered in the organic phase, DNA in the interphase, and RNA in the aqueous phase. Subsequently, the RNA can be precipitated from the aqueous phase.

The acid phenol/guanidine reagent was covered in the US under the protection of the US5346994, whose term should have ended in 2006. The European

**Table 6** Key IP rights protecting nucleic acid extraction techniques

| Company | Technology | End of 20 year term | Key IP right US | Key IP right EP |
|---------|-----------|--------|--------|--------|
| BioMerieux | Boom principle | 2010 | US5234809 | EP0389063 |
| Chomcynski, Piotr | Acid phenol/Guanidine mixture for nucleic acid preparation | 2006 | US5346994 | EP0554034 2013 |
| Amersham now GE Healthcare | Nucleic acid separation using magnetic beads | 2014 | US5523231 | EP0515484 |
| Dynal | Monodisperse polymer magnetic particles with different sizes | 2015 | US7173124 | EP0796327 |
| Dynal | Oligonucleotide-linked particles for specific nucleic acid separation | 2014 | US5512439 | EP0446260 2009 |

counterpart of the US5346994, the EP0554034, is still in force and will probably be so until 2013.

Apart from laborious and time-consuming traditional methods, alternative separation techniques have been developed for example using the capacity of DNA to bind to silica in the presence of high concentrations of a chaotropic salt (Boom principle). This method can be uniformly applied to all kinds of samples, such as whole blood, blood serum or plasma, sputum, sperm, feces, saliva, tissues and cell cultures, foods products, vegetable material, urine, tissue cells, body fluids, and biological material possibly infected with virus and bacteria. The patents US5234809 and EP0389063 covering the Boom principle are owned by Akzo Nobel and licensed to Organon Teknik, which was acquired by BioMerieux in 2001. The grant of the European patent was opposed by Eppendorf AG and Abbott Laboratories, but the patent was upheld. Thus, these patents in Europe gave Organon Teknik, i.e., BioMerieux, the right to enforce payment of royalties on both home-made and commercial nucleic acid purification systems based on the use of silica plus chaotropic salts until March 2010.

In a different approach, specifically functionalized magnetic particles were developed during the last few years. Together with an appropriate buffer system, they allow for quick and efficient purification directly from crude cell extracts avoiding centrifugation steps. In addition, the new approach provides for an easy automation of the entire process and the isolation of nucleic acids from larger sample volumes. Nucleic acid separation using magnetic beads is described in US5681946, US5523231, and EP0515484. All three patents were originally assigned to Amersham, which GE acquired in 2004. Their term will probably last until 2014 in the US and 2011 in Europe. In December 2009, GE Healthcare UK filed a suit against Beckman Coulter Genomics, in the US alleging that the firm was infringing the US5681946 and US5523231.[13]

---

[13] Berensmeier (2006).

The drawback of the techniques using magnetic beads is, that the approach is not nucleic-acid-specific, i.e., the magnetic beads adsorb other bio-substances in parallel.

This drawback is overcome by some of the several patents concerning nucleic acid separation granted to Dynal now owned by Invitrogen. These cover mono-disperse polymer magnetic particles with different sizes (coefficient of variation less than 5%) (EP0796327), which are sold with a polystyrene matrix under the name of Dynabeads®. The small-size distribution ensures reproducible separation properties. Protocols for nucleic acid separation with oligonucleotide-linked particles for specific nucleic acid separation are described in US5512439. Table 6 shows key IP rights protecting nucleic acid extraction techniques.

### 3.6.2 Methods for Formalin-Fixed Paraffin-Embedded Tissue Samples

Most of the methods described above can only reliably be used with fresh or frozen samples, hence they are not suitable for routine clinical practice especially in cancer research where samples are usually preserved through treatment with formaldehyde and paraffin. Thus, alternative methods were developed for isolation of nucleic acid from formalin-fixed paraffin-embedded tissue samples.

One of the companies involved in the research is Response Genetics, originally, Bio Type, Inc., which owns the patents US6248535 and EP1242594 covering a method for the isolation of RNA from formalin-fixed, paraffin-embedded (FFPE) tissue specimens. In this method, the tissue sample is first deparaffinized and further homogenized in a solution comprising a chaotropic agent, for example, guanidinium isothiocyanate. The homogenate is thereafter heated to about 100°C in a solution with a chaotropic agent. RNA is then recovered from the solution by phenol–chloroform extraction.[14]

Moreover, several other techniques for isolation of nucleic acid from FFPE samples have been devised, among them also the method of EP1777291, in which a formalin-fixed, paraffin-embedded biological tissue sample is de-paraffinized, contacted with a solution containing proteinase K, heated and afterwards treated again with proteinase K to significantly improve the recovery of RNA.

In order to further simplify the isolation of nucleic acid from FFPE samples the techniques utilizing functionalized magnetic particles were adapted to function with formalin-fixed, paraffin-embedded samples.

For instance Promega has launched the Maxwell® 16 FFPE Tissue LEV DNA Purification System, which facilitates purification of genomic DNA from formalin-fixed paraffin-embedded (FFPE) tissue using silica magnetic particles. The method is covered by US6027945, US6673631 and EP0895546.

In addition, Siemens has filed a patent application EP2288701 disclosing a method for filtering nucleic acids under non-chaotropic conditions, using very

---

[14] Ladanyi (2005).

**Table 7** Further IP rights protecting nucleic acid extraction techniques

| Company | Technology | End of 20 year term | Key IP right US | Key IP right EP |
|---|---|---|---|---|
| Response Genetics | Isolation of RNA from FFPE samples using a chaotropic agent | 2019 | US6248535 | EP1242594 |
| Siemens Healthcare Diagnostics | Method for filtering nucleic acids in particular form fixed tissue | 2028 | US2011092691 | EP2288701 |
| Pangaea Biotech | Isolation of RNA from FFPE samples using an additional proteinase K step | 2016 | US2009264641 | EP1777291 |
| Promega | Method for automated purification of genomic DNA from FFPE tissue using silica magnetic particles | 2020 | US6027945 | EP0895546 |

small (100 nm) magnetic particles with a very high iron content. Table 7 shows further IP rights protecting nucleic acid extraction techniques.

## 3.7 Recombinant DNA and Molecular Cloning

The development of recombinant DNA molecular cloning techniques have enabled the study of disease genes and their function, thus providing valuable insights for molecular diagnostic tests.

### 3.7.1 Recombinant DNA

Recombinant DNA is generated via introducing relevant DNA into an existing organismal DNA, such as the plasmids of bacteria, to code for or alter different traits for a specific purpose, such as antibiotic resistance. It differs from genetic recombination, in that it does not occur through processes within the cell, but is engineered.

The first patent application was filed by Stanford University in November 1974. This original patent application claimed both the process of making recombinant DNA and any products that resulted from using that process. The application was subsequently divided into a process patent application and two divisional product patent applications. The original patent application was abandoned. In December 1980, the process patent US4740470 for making molecular chimeras was issued. The other two related patents cover proteins produced using recombinant prokaryote DNA (US4237224) and proteins from recombinant eukaryote DNA (US4468464).

Table 8 Key IP rights for the protection of recombinant DNA techniques

| Company | Technology | End of 20 year term | Key IP right US | Key IP right EP |
|---------|-----------|------|------|------|
| Stanford and UCSF | Process patent for making molecular chimeras | 2005 | US4740470 | None |
| Stanford and UCSF | Proteins produced using recombinant prokaryote DNA | 1999 | US4237224 | None |
| Stanford and UCSF | Proteins from recombinant eukaryote DNA | 1998 | US4468464 | None |

The first licensee signed agreements with Stanford University on December 15, 1981. Although profit was not a primary motive, by the end of 2001, the patents had generated $255 million in licensing revenues, from licenses granted to a total of 468 companies. In this context, it should be remembered that the decision to negotiate nonexclusive rather than exclusive licenses was critical to the industry. If the technology had been licensed exclusively to one company and the entire recombinant DNA industry had been controlled by that company, the industry might never have developed. Alternatively, major pharmaceutical firms might have been motivated to commit their resources to challenging the validity of the patent.[15] Table 8 shows key IP rights for the protection of recombinant DNA techniques.

## 3.7.2 Molecular Cloning

Molecular cloning refers to the procedure of isolating a defined DNA sequence and obtaining multiple copies of it in vivo. In the classical restriction and ligation cloning protocols, cloning of any DNA fragment essentially involves the four steps of DNA fragmentation with restriction endonucleases, ligation of DNA fragments to a vector, transfection, and screening/selection.

Several methods to achieve molecular cloning are available today among them recombinase-based cloning. It works by inserting the special DNA fragment of interest into a special area of target DNA through interchange of the relevant DNA fragments. As this is a one-step reaction it is simple and efficient thus facilitating high throughput or automatic cloning and/or subcloning. One of the currently popular recombinase-based systems is marketed under the name Gateway Technology by Invitrogen under the license of Life Technologies.

Moreover, there are patents in place for nuclear transfer cloning also termed Somatic Cell Nuclear Transfer (SCNT). It is a laboratory technique for creating a clonial embryo, using an ovum with a donor nucleus. The possibility of producing live offspring by somatic cell NT carries potential applications in animal

---

[15] Bera (2009).

husbandry, biotechnology, transgenic and pharmaceutical production, biomedical research, and the preservation of endangered species. Two important patent families cover the use of nuclear transfer cloning. The first one including US5945577 and EP1015572 covers clones of any non-human mammalian species, generated from any adult or fetal somatic cell, during any phase of the cell growth cycle except quiescence. Using the proprietary technique, the first cloned transgenic cows produced from genetically altered bovine somatic cells, George and Charlie, were born in January 1998. The patents were issued to the University of Massachusetts, which licensed them exclusively to Advanced Cell Technology (ACT).

The other patent family comprising US6147276 and EP0849990 covers nuclear transfer technology using donor nuclei from quiescent cells. This method was developed by Keith Campbell and Ian Wilmut at the Roslin Institute in Edinburgh and was employed to clone the first animal, the famous sheep Dolly. The same technology has subsequently been used successfully by a number of other research groups to produce cloned mammals. The patents and the underlying technologies are licensed from Roslin to Geron for a wide range of applications, including use in human regenerative medicine and animal cloning. The license rights were obtained as part of Geron's acquisition of Roslin Bio-Med, Ltd. (now Geron Bio-Med, Ltd.).

In 2004, Geron requested that the U.S. Patent Office declare interferences between some of Geron's pending nuclear transfer patent applications and ACT patents US5945577 and US6234970 because, in Geron's view, the technology claimed in those patents was first invented at the Roslin Institute and was covered by the patent portfolio licensed to Geron. In late 2004 and again in early 2005, the Board of Patent Appeals and Interferences of the U.S. Patent Office ruled in favor of Geron. ACT, immediately filed appeals, naming Roslin, Geron and Exeter Life Sciences (Exeter), also a Roslin licensee, as defendants. In mid-2005, Geron and Exeter established a joint venture company, Start Licensing, to manage and license intellectual property rights for animal reproductive technologies, and exclusively licensed their rights in Roslin patent applications to Start in the nonhuman animal field. In 2006, it was announced that the patent dispute had been settled. Under the terms of the settlement agreement, Start was subjected to award ACT an initial payment of $500,000 and milestone payments of up to $750,000. Start, Geron, Exeter, and Roslin each further agreed not to sue ACT or UMass under the involved Roslin patent applications. In exchange, ACT and UMass dismissed their appeals, transferred control of related UMass patents and patent applications to Start in the non-human animal field, and ACT paid certain legal fees. Under the terms of the settlement agreement, ACT retained its rights under the UMass patents in the human field. This settlement resolves the parties' various patent rights with respect to nuclear transfer cloning in the non-human field.

Meanwhile, Infigen, which owns a suite of patents on the basic techniques of nuclear transfer, awarded before the Roslin team demonstrated that it is possible to clone mammals from differentiated cells, sued ACT in 1999 for breaching two US patents on cow cloning, one US5096822 covering a specific culture medium, the other US6147276 a method for activating bovine eggs after transferring the donor

**Table 9** Key IP rights protecting molecular cloning techniques

| Company | Technology | End of 20 year term | Key IP right US | Key IP right EP |
|---|---|---|---|---|
| Life Technology licensed to Invitrogen | Gateway recombinase-based cloning | 2016 | US5888732 | EP0937098 |
| Expression Technologies | Synthesized plasmid | 2022 | US20030157661 | None |
| New England biolabs | Examples of restriction endonucleases | 2021 | US6596524 | EP1298212 |
| Geron Corp | Nuclear transfer cloning | 2016 | US6147276 | EP0849990 |
| University of Massachusetts Licensed to ACT | Clones of any non-human mammalian species, generated from any adult or fetal somatic cell, during any phase of the cell growth cycle except quiescence | 2017 | US5945577 | EP1015572 |
| Infigen | Culture medium | 2010 | US5096822 | None |
| Infigen | Oocyte activation | 2013 | US5496720 | None |

nucleus. Infigen also claimed that a researcher, who had once worked for Infigen, stole its trade secrets. That complaint was rejected, but in June 1999, the U.S. District Court in Wisconsin ruled that ACT had indeed infringed Infigen's patents. After this ruling the two companies came to a confidential settlement. Table 9 shows key IP rights protecting molecular cloning techniques.

## 3.8 DNA Methylation

DNA methylation is a natural and tightly controlled biological process that serves the regulation of genes and the stability of the genome. In DNA methylation cytosine, one of the four bases in DNA, is modified by the covalent addition of a methyl group. Every cell type has its unique DNA methylation "fingerprint" that changes in various normal biological processes and in many diseases, in particular cancer. Thus, DNA methylation has been proposed as rich source for highly specific biomarkers for organ-specific disease diagnosis, classification and prediction for therapeutic intervention.

One of the companies with a patent portfolio in DNA methylation is Epigenomics. The patents cover DNA methylation biomarkers in numerous cancer indications and other diseases as well as technologies for their discovery and testing in research and diagnostic applications. For example, in May 2009 Epigenomics was granted two patents for the PITX2 DNA methylation biomarker (mPITX2). EP1831399, covers very broadly the use of the mPITX2 biomarker in the prognosis of prostate cancer, while EP1554407 claims the use of mPITX2 in the prediction of the response of breast cancer patients to antihormonal therapy. The mPITX2 biomarker is at the core of Epigenomics' prostate cancer prognosis test that indicates the risk of early disease recurrence following a surgical removal of the cancerous prostate. Furthermore, PITX2 methylation status could also be used in the rational design of clinical trials and improve testing of new therapeutic regimens.[16]

A cooperation between Epigenomics and Qiagen led to the launch of the EpiTect™ Bisulfite Kit, by Qiagen in the spring of 2006. The kit facilitates the complex and time-consuming step of bisulfite treatment of DNA in DNA methylation analysis. Treatment of DNA with bisulfite converts cytosine residues to uracil, but leaves 5-methylcytosine residues unaffected. Thus, bisulfite treatment introduces specific changes in the DNA sequence that depend on the methylation status of individual cytosine residues, yielding single-nucleotide resolution information about the methylation status of a segment of DNA. Various assays for instance PCR can be performed on the altered sequence to retrieve this information. The kit is inter alia covered by the US5786146 and the EP0954608, which Qiagen has licensed from the John Hopkins University.

---

[16] Press release of May 20, 2009.

**Table 10** Key IP rights protecting methylation detection techniques

| Company | Technology | End of 20 year term | Key IP right US | Key IP right EP |
|---|---|---|---|---|
| John Hopkins University licensed to Qiagen | Methylation specific PCR | At least 2017 | US5786146 | EP0954608 |
| Epigenomics | mPITX2 in prostate cancer prognosis | 2025 | US20090197250 | EP1831399 |
| Epigenomics | mPITX2 in breast cancer therapy | 2023 | US2006121467 | EP1554407 |

Furthermore, Epigenomics and Roche started a cooperation in 2002 to develop a range of molecular diagnostic and pharmacogenomic products for the early detection of cancer. This cooperation was based on the cancer marker Septin 9, protected by Epignomics EP1721992 and US7749702, whose overexpression has also been demonstrated in a number of tumor tissues. However, Roche ended the cooperation in 2006. Table 10 shows key IP rights protecting methylation detection techniques.

# 4 Key Applications

## 4.1 Infectious Diseases

An example of an infectious disease-related patent is the US7611704 which was granted to the University of Texas System and is exclusively licensed to Peregrine Pharmaceuticals. It includes broad claims covering anti-viral uses of phosphatidylserine (PS)-targeting antibodies including Peregrine's lead clinical compound, bavituximab. PS, a lipid molecule normally found only on the inside of cell membranes, becomes exposed on the outside of the membranes of virally infected cells. A rapidly growing body of published scientific research confirms that exposed PS is directly involved in the pathogenesis of many serious infectious diseases. Exposed PS enables viruses to evade immune recognition and dampens the body's normal responses to infection. By masking the exposed PS, PS-targeting antibodies are believed to block these effects, allowing the body to develop a robust immune response to the pathogen. Targeting PS thus provides a broad platform for treating viral infections. Because the PS target is host-derived rather than pathogen-derived, PS-targeting antibodies are also expected to be much less susceptible to the viral genomic mutations that lead to anti-viral drug resistance.

The new patent covers compositions and methods of treating virus infections using bavituximab and similar antibodies, either alone or as immunoconjugates attached to anti-viral agents, as well as in combination with other anti-viral agents. The breadth of the claims in this patent is especially noteworthy, since it includes methods for treating all viruses in humans and animals. The therapeutic potential of these antibodies is supported by data in a landmark study published in the December 2008 edition of *Nature Medicine* showing that bavituximab and other PS-targeting antibodies have the potential to cure lethal virus infections across a broad range of virus families in animal disease models.[17]

A technique useful in the infectious disease segment is the FullVelocity technology from Stratagene. It is a high-speed reagent system for quantitative PCR and qRT-PCR that delivers sensitive, specific, and reproducible results with significantly shorter run times than competing methods. The technology can be used to identify infectious diseases, cancer, genetic diseases, and drug sensitivities. Stratagene has five US issued patents US6350580, US6528254, US6548250, US6589743, and US6893819 on the technology with additional patents pending. The company has entered into a collaborative agreement with Focus Diagnostics over the patents. Under the terms of the agreement Stratagene has consented to license its Full Velocity™ nucleic acid amplification technology and associated know-how to Focus for the development of molecular diagnostic products.[18]

Another company developing diagnostic tests and technology for the detection of infectious diseases is Chembio Diagnostic Systems, which received US7189522 in 2007 for its Dual Path Platform (DPP®) a point of care diagnostic product. Several new tests have already been developed on DPP® including a new oral fluid HIV 1/2 test, a new five-band Point of Care (POC) confirmatory test for HIV 1/2, and a combination screening and confirmation test for Syphilis. The HIV tests have been selected in the national testing protocols of several countries in Africa, Asia and Latin America. Moreover, Chembio works with organizations such as the United States National Institutes of Health (NIH) and the Infectious Disease Research Institute (IDRI) in order to develop rapid POC tests, all on DPP®, for Tuberculosis, Malaria, Leprosy, Leishmaniasis, Leptospirosis, Syphilis, HIV, Influenza, Hepatitis C, and other conditions in humans and animals.

Furthermore, Chiron Corporation, which was acquired by Novartis in 2006, has a large portfolio in hepatitis C virus patents. Chronic HVC infections can progress to scarring of the liver with consequences such as liver failure or liver cancer. In the late 1980s, Chiron Scientists together with Scientists of the Center for Disease Control and Prevention (CDC) of the United States were the first to clone HVC. Chiron filed for numerous patents on the virus and its diagnosis, the EP0318216, being one of the earliest. A competing patent application by the CDC was dropped in 1990 after Chiron paid a substantial sum to the CDC. In 1998, Chiron filed patent infringement suits in Europe, Japan, and the US against Roche over its hepatitis C (HCV) products. A settlement was reached when Roche agreed to buy the global semi-exclusive nucleic acid test (NAT) patents for HCV and HIV from Chiron. In 2004, Chiron held 100 patents in 20 countries related to hepatitis C and had successfully sued many companies for infringement. The dominance of

---

[17] Press release of November 5, 2009.
[18] Press release of December 5, 2005.

**Table 11** Key IP rights covering techniques for the diagnosis of infectious diseases

| Company | Technology | End of 20 year term | Key IP right US | Key IP right EP |
|---|---|---|---|---|
| Stratagene | Full Velocity™ nucleic acid amplification technology | 2019 | US6350580 | EP1228242 |
| The University of Texas System Peregrine Pharmaceuticals | Anti-viral uses of phosphatidylserine (PS)-targeting antibodies | 2023 | US7611704 | EP1537146 |
| Chembio Diagnostic Systems | Dual Path Platform point of care | 2025 | US7189522 | EP1856503 |
| Chiron | Hepatitis C virus | 2008 | US5712087 | EP0318216 |

Chirons patent portfolio is such that literally any company that develops a new drug targeting hepatitis C (such as a protease inhibitor), or a diagnostic test to detect and measure HCV (viral load; test for screening the blood supply), needs to license Chiron's patents. Table 11 shows key IP rights covering techniques for the diagnosis of infectious diseases.

## 4.2 Cancer-Related Applications

As mentioned in the introduction, many of the best known molecular diagnostics patents stem from the field of disease gene patents. For example, a recent search in the patent (application) database qpat revealed more than 2400 patents and patent applications for cancer-related applications of molecular diagnostic patents ((MOLECUL + AND DIAGNOS +) AND ((CANCER OR TUMOR OR MALIGN + OR NEOPLAST +)) AND (MRNA OR TRANSCRIPT OR (EXPRESSION AND PROFILE)) October 2011). Of these patents, 32 % were listed in the class C12Q-001/68M6B which inter alia comprises mutations or genetic engineering indicating the relation of the patents to the basic techniques discussed in the previous chapter. Moreover, 5.2 % of the patents are listed in class G01 N-033/574C4 and thus are dealing with chemical analysis of biological material.

### 4.2.1 BRCA

The most famous cancer-related patents are probably the patents for the BRCA1 and BRCA2, genes, two critical genes implicated in the development of breast and ovarian cancer. Women with the mutant form of the BRCA genes, which are implicated in up to 10% of breast cancers, usually develop the disease at an early age. Inherited mutations can reportedly increase a woman's risk of breast cancer from 9 to 80%, and the risk of ovarian cancer from 1 to 65%.

While the BRCA genes are now, at least in Europe private property, their initial discovery was due to the efforts of publicly funded scientists collaborating on an international basis. Breast cancer genetics were first developed in the 1980s with research groups based in the US, UK, France, Japan, and Canada. In 1988, the International Breast Cancer Linkage Group (BCLG) was formed to bring together researchers in the field. In 1990, researchers at the University of California, Berkeley announced the discovery of the BRCA1 gene and indicated its association with breast cancer. The BCLG then conducted tests with 214 families to prove the hereditary character of the condition.

At the same time, a team based at the University of Utah's Centre of Genetic Epidemiology was also working on the identification of cancer genes. Scientists used a genealogical database of 200,000 Mormons that was checked against Utah's cancer registry. The research team founded the firm Myriad Genetics and used its exclusive access to the Utah and Mormon databases to attract investment from the giant US pharmaceutical company Eli Lilly. Myriad published its findings in 1994, and in 1995 obtained a patent for the sequence of the BRCA1 gene and methods of detecting its mutations.

Also in 1994, researchers from the University of Utah and a British group identified a second gene, BRCA2. The sequence for the gene was published in 1995 by the British group in collaboration with 40 scientists in six different countries. Myriad then published its own sequence a few months later, claiming the published sequence from the British team was incomplete. By 2000, it had obtained the patents to the second gene. Myriad secured a total of nine patents, giving it control of BRCA1 and 2, including any diagnostic tests based on the genes. Myriad has pursued its patent around the world. For instance, the company has fought protracted legal battles in the European Union courts to have its "property" recognized.

Until the companies owning the patent rights in the different countries or licensing the rights from Myriad Genetics started to enforce the patents, screening could be carried out by public laboratories without charge. Therefore, there was a public outcry when it became known that the screening would be made more expensive due to royalty fees. This was especially the case, since the tests themselves are not difficult procedures, any laboratory set up for DNA diagnostic testing would be competent to perform them, i.e., there is no specific equipment or instrument or special reagent kit or product for sale. The product is performing the test itself.[19]

While the European patents listed below are upheld in amended form, the seven US patents were invalidated on March 29, 2010. Specifically, District Judge Sweet in the case of *Association for Molecular Pathology* Et al. *v. USPTO and Myriad Genetics* ruled that according to his opinion, genes do not constitute patentable subject matter. The key basis of the judgment was that the BRCA1 and BRCA2

---

[19] Terry (2006).

**Table 12** Key IP right BRCA diagnosis

| Company | Technology | End of 20 year term | Key IP right US | Key IP right EP |
|---------|-----------|---------------------|-----------------|-----------------|
| Myriad | BRCA1/2 | At least 2015 | US5693473 | EP0699754 |

genes constituted a "product of nature", that was not significantly different from their original form. Judge Sweet noted that *were the isolated BRCA1/2 sequences different in any significant way, the entire point of their use—the production of BRCA1/2 proteins—would be undermined.* The claims in suit were declared invalid on the basis that *as determined above, the patents issued by the USPTO are directed to a law of nature and were therefore improperly granted.* Table 12 shows the respective key IP.

### 4.2.2 Applications Based on Knowledge Gained by Molecular Diagnostics

These two examples also demonstrate how a key cancer-related diagnostic information was first developed in the public sector, while the firms wishing to market the drug subsequently obtained the rights to use this information and proceeded to coordinate product and test development. They took the lead in organizing a collaboration with a diagnostics firm to make sure that the necessary tests were on the market. This may well be the future model. The drug firms have superior access to clinical trial data and significantly greater resources than diagnostic firms, although few have the in-house expertise to develop diagnostics themselves. There is thus an incentive to develop a relevant diagnostic test, even without a patent and the absence of a patent may even make it easier if different drug manufacturers need access to the same genetic mutations.

The first example concerns the monoclonal antibody (mAB) Herceptin. It was developed by Genentech and marketed by Roche, for a category of breast cancers in which the *ERBB2* oncogene is overexpressed. About 35% of women have breast cancers in this category, and one can test for this expression in several ways. Patents related to *ERBB2* have been widely sought. The key gene sequence, however, was identified quite early, the first relevant scientific article was published in 1987 by a team, which included several academic scientists as well as a Genentech scientist. The University of California obtained a patent (US4968603) on the use of *ERBB2* to determine disease status in 1990. Genentech subsequently developed a mAb-based drug to be used on those patients overexpressing the gene. It entered partnerships to ensure availability of the diagnostic test for *ERBB2*. Genentech joined with Dako, Vysis, now a subsidiary of Abbott Laboratories which markets an *ERBB2* FISH test and Ventana Medical Systems, which in 2002 obtained a patent on a fluorescent in situ hybridization (FISH) assay for *ERBB2* (US6358682). Genentech has also obtained a patent on a particular procedure using FISH, which covers determining *ERBB2* expression (US6573043).

**Table 13** Examples for IP rights based on knowledge gained by molecular diagnostics

| Company | Technology | End of 20 year term | Key IP right US | Key IP right EP |
|---|---|---|---|---|
| University of California Licensed to Genentech | Use of ERBB2 to determine breast cancers disease status | 2006 | US4968603 | None |
| Ventana Medical Systems | Fluorescent in situ hybridization (FISH) assay for ERBB2 | 2019 | US6358682 | None |
| Genentech (owed by Roche) | Another FISH for determining ERBB2 expression | 2018 | US6573043 | None |
| MIT Repligen | Genetic elements that increase protein production | 2004 | US4663281 | None |
| Yeda (Sanofi-Aventis) licensed to ImClone | mAb specific to human epidermal growth factor receptor & therapeutic use | 2018 | US6217866 | EP0359282 2009 |

The second example, Erbitux, is ImClone's mAb targeted against Epidermal Growth Factor Receptor (EGFR) for the treatment of metastatic colorectal cancer. Originally, Imclone marketed Erbitux, first known as C255, based on US6217866, which was granted in 2001. At roughly the same time in early 2004 that the US Food and Drug Administration (FDA) approved the drug Erbitux, it also approved a test, EGFRpharmDx, developed by DAKO to determine which patients would be most likely to benefit from Erbitux. According to a Dako press release, this product was used in the Erbitux clinical trials and its approval was sought 'in parallel' with that of Erbitux.[20]

However in 2003, Yeda Research Development, the commercial arm of the Weizman Institute of Science in Israel, took Imclone to court over US6217866. The reason being that Erbitux was originally developed in the 1980s by three scientists at the Weizman Institute, one of which, who later worked at a predessesor company to Aventis, licensed Erbitux to Imclone. In 2007, after various litigations in different countries the matter was finally settled. Under the terms of the agreement Yeda was declared the sole owner of the patent in the US while Yeda and Sanofi-Aventis became co-owners of the patents foreign counterparts. ImClone further agreed to pay royalties to Yeda.

In addition, ImClone due to its manufacture and sale of Erbitux, was accused of infringement of US4663281 by Repligen and the Massachusetts Institute of Technology (MIT) in 2004. Repligen and MIT alleged that the cell line that ImClone used to produce Erbitux employed key technology that is claimed in US4663281. They further alleged that the cell line was created under contract for the National Cancer Institute (NCI) by a predecessor to Repligen and subsequently transferred from the NCI to ImClone for use in research and development only. In 2006, it was ruled that

---

[20] Barton (2006).

neither the transfer to the NCI by Repligen's predecessor nor the subsequent transfer to ImClone by the NCI exhausted the proprietary rights of Repligen and MIT. The final settlement was reached between the parties in 2007, including the provision for ImClone to make a payment of $65 million to co-plaintiffs.[21] Table 13 shows examples for IP rights based on knowledge gained by molecular diagnostics.

## 4.3 Further Important Patents

Moreover, BioCurex Inc. in 2003 announces the grant of US6514685 on the cancer marker known as RECAF, which is found on malignant cells from a variety of cancer types but is absent in most normal or benign cells. The patent covers the technologies used in several kits to be released for cancer diagnosis on blood samples by BioCurex (press release 11.03.2003 at http://www.biocurex.com). On the 8th of September, 2011 it was announced the use of the BioCurex RECAF blood test can prevent 70% of unnecessary prostate biopsies.[22]

Furthermore DxS, owned by Qiagen, has developed a set of molecular diagnostic assays, which allow physicians in oncology to predict patients' responses to certain treatments in order to make cancer therapies more effective and safer. The currently marketed portfolio spans seven real-time PCR tests including a test for the mutation status of the oncogene K-RAS, termed Thera-Screen, which is indicative of the successful treatment of patients suffering from metastatic colorectal cancer (mCRC) with EGFR inhibitors.

However, in February 2010, a unit of Roche Holding AG sued DxS, accusing it of trying to break a distribution agreement following its acquisition by Qiagen. In this context it has to be noted that Roche was in danger of losing the right to sell the two TheraScreen tests, if DxS got out of the contested agreement, which was signed in May 2008. While Roche was of the opinion that DxS is breaching the distributor agreement, DxS argued that Roche failed to assist DxS in the development of software. However, in Roche's opinion this was a pretext because it had no such obligation under the distributor agreement. In the end the two companies managed to settle their dispute outside of a court. Under the terms of the settlement, Roche has maintained its existing distribution rights to DxS' TheraScreen KRAS and TheraScreen EGFR assays. It also retains rights to distribute future versions of these products under certain conditions. Additionally, the firm has been granted an option to extend the term of the distribution agreement for the TheraScreen EGFR assay, beyond 2011 when it currently ends. Qiagen, in return, retains the rights to distribute the KRAS and EGFR assays under its own label.

---

[21] Press release Repligen 08.11.2007.
[22] Press release 08.09.2011 at http://www.biocurex.com.

**Table 14** Further important IP rights discussed herein

| Company | Technology | End of 20 year term | Key IP right US | Key IP right EP |
|---|---|---|---|---|
| Epigenomics | Method for detection of cytosine methylation | 2022 | US7229759 | EP1370691 |
| Epigenomics | Septin 9 | 2026 | US7749702 | EP1721992 |
| DxS | Scorpion-primers | 2018 | US6326145 | EP1088102 |
| Genomic Health | Oncotype DX$^{TM}$ assay | At least 2023 | US7723033 | EP1641810 |
| | | | US7081340 | EP1488007 |
| | | | | EP1918386 |
| Digene Now Qiagen | Assessment of HPV-Related Disease | 2018 | US6355424 | EP1038022 |
| Owed by Digene | HPV type 52 DNA sequence | 2017 | US5643715 | EP0370625 |

It also has exclusive distribution rights to all other assays including future assays developed and manufactured by its DxS subsidiary.[23]

A second kit from DxS is the BCR-ABL T315I for the detection of the T315I Mutation in the BCR-ABL Fusion Gene. The scorpion-primer technology used in both kits is protected under US6326145 and EP1088102.

In addition, Genomic Health introduced its Oncotype DX$^{TM}$ assay, which analyses the expression levels of 21 genes in an effort to predict the probability of recurrence in early stage invasive breast cancer in 2004. The information helps physicians to decide whether or not to prescribe chemotherapy. The assay is protected by US7723033, US70056674, as well as US7081340 and their European counterparts.

Moreover, Digene Corporation (now belonging to Qiagen) owns a set of patents relating to the human papillomavirus (HPV). While most infections with HPV, even with high risk subtypes do not lead to a serious disease, HPV is a cause of nearly all cases of cervical cancer. Therefore, Digene developed the HPV HC2 DNA Test. It is an in vitro nucleic acid hybridization assay with signal-amplification using microplate chemiluminescence for the qualitative detection of 18 types of human papillomavirus (HPV) DNA in cervical specimens. The Digene HC2 HPV DNA test was approved by the FDA and is used together with the conventional cervical Papanicolaou (Pap) test, which detects abnormal cells. Table 14 shows further important IP rights discussed herein.

---

[23] Roche press release of 26 May 2010.

**Table 15** Leica's Key IP rights covering Laser Microdissection

| Company | Technology | End of 20 year term | Key IP right US | Key IP right EP |
|---|---|---|---|---|
| P.A.L.M. subsidiary of Carl Zeiss | Laser microdissection and laser pressure catapulting | 2017 | US5998129 | EP0879408 |

## 4.4 Laser Microdissection

Laser microdissection is an extraction process to dissect specific tissue for analysis or research. Specifically, a laser is coupled into a microscope and focuses onto the tissue. By movement of the laser or the stage the focus follows a trajectory which is predefined by the user. This trajectory, a so-called *Element*, is then cut out and separated from the adjacent tissue. After the cutting process, an extraction process has to follow. One available method is the laser microdissection and laser pressure catapulting (LMPC) technique from P.A.L.M. Microlaser Technologies GmbH, a subsidiary of Carl Zeiss MicroImaging. In this system the sample, following microdissection, is directly catapulted into an appropriate collection device. As the entire process works without any mechanical contact, it enables pure sample retrieval from a morphologically defined origin without cross contamination. Therefore, this technique results in samples that can be directly employed in various downstream applications, such as single-cell mRNA-extraction or different PCR methods. The LMPC method is covered by EP0879408 and DE19603996.

In September 2000, P.A.L.M. Microlaser Technologies AG on the basis of the EP0879408 filed patent infringement proceedings against Leica Microsystems, because Leica was selling a system, in which with the help of a laser beam biological objects were cut from a preparation on an object carrier leaving only a bridge and then, by means of a laser shot directed toward the bridge and cutting through the bridge, are catapulted toward a collection device. In P.A.L.M.'s opinion this made unlawful use of the teaching protected by the European patent. On April 6, 2005, Leica was sentenced for infringement of P.A.L.M. patent EP0879408.

However, the dispute was not finally settled until 2007, when both parties agreed that Leica would have a right of continuation for the systems delivered in the past, including continued use and maintenance. Furthermore, Leica is authorized to offer and distribute laser micro-dissection systems with the alternative cutting technique used in the current system. P.A.L.M., on the other hand, continues to maintain sole authorization to manufacture and distribute laser microdissection systems with the so-called "catapult procedure" in accordance with the above-mentioned P.A.L.M. patents. Both parties have agreed to maintain secrecy regarding the further stipulations of the settlement.[24] Table 15 shows P.A.L.M.'s respective key IP.

---

[24] Press release by P.A.L.M. 28.06.2007.

**Table 16** Key IP rights protecting molecular forensics methods

| Company | Technology | End of 20 year term | Key IP right US | Key IP right EP |
|---|---|---|---|---|
| Max-Planck-Gesellschaft Licensed to Promega | STR analysis | 2009/2018 | US5766847 | EP0438512 |
| Cetus Licensed to Roche | HLA typing | 2003 | US4582788 | EP0084796 |

## 4.5 Forensics

Forensics, especially forensic DNA analysis, takes advantage of many techniques of molecular diagnostics, mainly in order to use the uniqueness of an individual's DNA to answer forensic questions such as whether a suspect has been at a crime scene.

The most obvious technique employed in forensic analysis is of course PCR, apart from the general method covered by core patents as described above, several forensic specific techniques have been developed. The first commercial PCR kits used in forensic analysis was Cetus's "HLA DQ-alpha" and the subsequent "HLA DQAt" kit by Roche Molecular System. These kits rapidly amplified and typed four alleles at the HLA DQAt locus (a locus in the human leukocyte antigen system), while requiring very little sample. Comparison of different samples indicates whether they have the same HLA DQAt genotype and could potentially have come from the same source.

Another example of a patent covered technique employed in forensic analysis is the Short Tandem Repeat (STR) analysis offered by Promega. Promega was the first company to provide kits for STR analysis of single loci. It co-operated with the FBI and other crime labs in validating STR loci that would eventually be selected as the core loci for the Combined DNA Index System (CODIS), used for forensic DNA testing in North America. CODIS is a computer system that stores DNA profiles created by federal, state, and local crime laboratories in the US, with the ability to search the database to assist in the identification of suspects in crimes. Currently, a similar database is being set up in Germany. While the relevant core patents EP0438512 and US5766847 were originally filed by the Max-Planck-Gesellschaft, Promega has acquired an exclusive sublicense for a number of different fields of application.[25]

In addition, as discussed above Promega has launched the Maxwell® 16 FFPE Tissue LEV DNA Purification System for purification of genomic DNA from formalin-fixed paraffin-embedded (FFPE) tissue, which can also be employed in forensic analysis. Table 16 shows key IP rights protecting molecular forensics methods

---

[25] http://www.analytica-world.com/news/d/47756/

**Table 17** Key IP right protecting PNA Technology

| Company | Technology | End of 20 year term | Key IP right US | Key IP right EP |
|---------|-----------|--------------------|-----------------|-----------------|
| Isis pharmaceuticals Inc. | PNA | 2017 | US6710164 | EP0960121 |

# 5 Outlook

In the wake of the human genome project more and more large public funded studies are organized, which bring together the knowledge and facilities of many different research institutions. An example of such a study is the EU project on diagnostic tools for improving disease detection, which will help to detect defective genes more quickly and at a lower cost. These tests will be validated using the cystic fibrosis gene, but will have potential applications for any genetic disease. The project will exploit the inherent advantages of a new kind of probe, PNA, or peptide nucleic acid, which is a laboratory-created model of DNA invented in Europe in the early 1990s and protected inter alia by the patents US6710164 and EP0960121. Because of the nature of its neutrally charged chemical backbone, the PNA–DNA bond is stronger than the DNA–DNA bond within normal probes. The project aims to build two new diagnostic systems for genetic testing based on PNA probes arranged in a microarray (or DNA chip) format, and improved capillary electrophoresis—a separation technique in which different biomolecules move at different speeds through an electrically charged capillary tube. Involved in the project are six partners, namely the Hôpital Henri Mondor (France), the Institute of Biocatalysis and Molecular Recognition (Italy), the Max-Planck-Institut für Molekulare Genetik (Germany), Innosense Srl (Italy), MEDWAY SA (Switzerland), and IMSTAR (France). Table 17 shows the respective key IP assigned to Isis.

Moreover, personalized medicine is an often-mentioned topic in connection with molecular diagnostics. Over the past century, medical care has centered on standards of care based on epidemiological studies of large cohorts. However, large cohort studies do not take into account the genetic variability of individuals within a population. Personalized medicine seeks to provide an objective basis for consideration of such individual differences. Traditionally, personalized medicine has been limited to the consideration of a patient's family history, social circumstances, environment, and behaviors in tailoring individual care.

Advances in a number of technologies including molecular diagnostics may allow for a greater degree of personalized medicine than the one currently available. Information about a patient's proteinaceous, genetic, and metabolic profile could be used to tailor medical care to that individual's needs. A key attribute of this medical model is the development of companion diagnostics, whereby molecular assays that measure levels of proteins, genes, or specific mutations are used to provide a specific therapy for an individual's condition by stratifying disease status, selecting the proper medication, and tailoring dosages to that patient's specific needs.

Additionally, such methods can be used to assess a patient's risk factor for a number of conditions and tailor individual preventative treatments.

First examples of successful personalized treatments already exist in the field of oncology in the form of the test for ERBB2 and EGFR proteins in breast, lung, and colorectal cancer patients as discussed above.

However, in most cases the diagnostic tools and possibilities are far greater than the existing therapies to cure the specific diagnosed disease subtype. Thus, it could happen that a cancer patient, for instance, will know the exact molecular signature of his tumor, but he will still be forced to contend himself with the standard therapy as nothing else is yet available or approved.

# 6 Considerations Concerning Patent Protection

## 6.1 Introduction

The scope of patent protection is determined by the patent claims. These can be directed at a physical entity, i.e., a product or a compound or a device. Such a claim is usually referred to as "product claim". Alternatively, a claim can protect an activity, i.e., a process or a method or a use. This type is thus termed "process claim" or "use claim".

In general, a patent which claims a physical entity per se (e.g., in the chemical field: a compound X), confers absolute protection upon such physical entity; that is, wherever it exists and whatever its context, and therefore for all its uses, whether known or unknown (decision G 2/88, OJ EPO 1990, 93, point five of the reasons). It follows that if a compound X is already known in the state of the art "for use as a dye", a later claim to the same compound X per se "for use as a catalyst" lacks novelty. This means that the indication of the use, purpose or function in a product claim is normally not seen as having a limiting effect on the scope of protection. The second inventor has to be contented with a "use" or "method" claim and the patent, if granted, will be dependent upon the prior compound X patent, if any. Only when a known chemical product (or composition), e.g. compound X, is found for the first time to have a medical application (e.g. for the cure of hepatitis), a so-called "first medical use" limited product claim can be put forward. Such a claim can read generally e.g.: "Compound X for use as a medicament", or specifically: "Compound X for use in treating hepatitis". If a patent is granted for such a claim, the patent will of course be dependent upon the prior compound X patent, if any. With the advent of the EPC 2000, it became possible under the new Article 54(5) to put forward a "second or further medical use" limited product claim when a second or further new (and inventive) specific medical use is found for the same compound X (e.g. "Compound X for use in the treatment of rheumatoid arthritis" or "Compound X for use in the treatment of hay fever"). If a patent is granted for such a claim, the patent will be dependent upon the prior compound X patent as well on the prior "first medical" use patent, if any.

## 6.2 Biological Compounds

For biological compounds such as DNA or proteins slightly different rules apply.

Under European patent law the use of a DNA sequence or a protein must be specified in the application. It is not sufficient to state for which protein a DNA encodes. Instead, also the function of the particular DNA sequence has to be disclosed.

During the process of transposition of the Directive 98/44/EC into the national laws of the EU member states, an ethical and political debate was initiated on the question whether patents on genes (DNAs) should be allowed according to "absolute product protection" or to "purpose bound protection". However, the Directive does not expressly mention the need to claim the industrial application of the gene sequence. Nevertheless, some national laws including Germany implementing the European Directive made it compulsory to specify the concrete function performed by the genetic sequence in the claims.

Moreover, according to a recent decision of the European Court of Justice[26] in a case concerning pesticides, absolute compound protection for DNA was denied, even though an appropriate claim had been granted. In this instant, the level of protection was reduced to the purpose bound scope of the claim. Thus, at the moment it seems that DNA patents are not patentable as chemical or biological compounds as such but only in the more limited context of their specific use and/or function.

## 7 Drafting Recommendations Concerning Molecular Diagnostics

In this chapter a brief overview is provided over the main points to consider when drafting a patent application covering molecular diagnostics.

## 7.1 Nucleic Acid Patents

As stated in C IV 5.4 of the guidelines for examination of the EPC it is necessary to disclose the industrial application of a nucleic acid in the patent application. A mere

---

[26] Decision of the European Court of justice in case C–428/08 (Monsanto Technology LLC vs Cefetra BV).

nucleic acid sequence without indication of a function is not a patentable invention (EU Dir. 98/44/EC, rec. 23). Moreover, in cases where a sequence or partial sequence of a gene is used to produce a protein or a part of a protein, it is necessary to specify which protein or part of a protein is produced and what function this protein or part of a protein performs.

Since the European patent office requires each patent application containing nucleic acid sequences to be accompanied by a separate sequence listing, the most straightforward way of claiming the sequences it to identify them by their number, e.g., primer GTCATGGTAC corresponds to sequence number X in the sequence listing and is referred to in the application as SEQ ID No. X.

When claiming a stretch of DNA sequence, i.e., a small molecule like a primer, a larger entity like a promoter element or a whole gene it will always be beneficial to not only claim the exact sequence but also fragments, variants, homologs, or derivatives thereof. Otherwise it would be very easy for a competitor to find a way to work around the claimed sequence.

A typical claim taking this consideration into account might read: "A nucleic acid molecule, selected from the group consisting of

(a) a nucleic acid molecule having the nucleotide sequence of SEQ ID NO: X, or a fragment, variant, homolog, or derivative thereof, and/or
(b) a nucleic acid molecule having a sequence identity of at least 70, preferably 95 % with any of the nucleic acid molecules of (a)."

When using this type of claim wording it is of uttermost importance to exactly define the terms "fragment, variant, homologue and derivative" and to specify with which method the sequence identity should be determined e.g. with NCBI Blastn, version XXX.

Furthermore, to allow for the fact that it is in theory sufficient to add a few nucleotides to the nucleic acid sequence in the claim above to evade the scope of protection of the claim, a nucleic acid molecule comprising the nucleic acid molecule presented as SEQ ID NO: X might be claimed.

Moreover, in order to claim multiple nucleic acid molecules together—with more than 15 claims each claim costs extra before the EPO—the wording, called Markush group, as shown below might be used.

A nucleic acid molecule, selected from the group consisting of

(a) the nucleic acid molecules presented as SEQ ID NO: X–X,
(b) a nucleic acid molecule comprising one of the nucleic acid molecule presented as SEQ ID NO: X–X,
(c) a nucleic acid molecule that is capable of hybridizing to any of the nucleic acid molecules of (a)–(b) under stringent conditions, and/or
(d) a complement of any of the nucleic acid molecules of (a)–(c).

Again it is of vital importance to define exactly when is meant with "capable of hybridizing under stringent conditions" and "complement" in the specification of the patent application.

## 7.2 Depositing a Cell Line

Deposition of a cell line may be an adequate way of specification in order to avoid sequencing errors and typographical errors, or to provide enabling information for features, which relate to post-translational modifications (i.e., unusual glycosylation patterns). The deposition process is subject to laws and bylaws provided by the respective patent legislations.

## 7.3 Claim Wording

Patents may be obtained for new products, particularly substances or compositions, for use in these methods of treatment or diagnosis. However, as European patents are not granted for methods for treatment of the human or animal body by surgery or therapy and diagnostic methods practiced on the human or animal body care needs to be taken with the claim wording.

According to point C IV 4.8 of the guidelines for examination of the EPC a claim in the form "Use of substance or composition X for the treatment of disease Y..." will be regarded as relating to a method for treatment explicitly excluded from patentability under Art. 53(c) and therefore will not be accepted.

However, the EBA in its decision G02/08—point 7.1. of the reasons—ruled that also the claim wording referred to as "swiss type" (shown below) is no longer acceptable:

Use of a substance or composition for preparing a medicament for therapeutic or prophylactic treatment and/or diagnosis of disease Y.

The exact format of an allowable claim was not given, but it probably runs along the lines of:

Compound X for (use in a method for) treating disease Y.

Furthermore, the most straightforward way to avoid that a method might be classified as diagnostic method and hence is considered not patentable, is to ensure that the outcome of the method is not attributed to a particular clinical picture (see point 2 above). It should be noted however, that this can only be done, if the inventive step does not lie in exactly this attribution since the examining division will then insist that this information will be included in the claim, in order to meet the requirement for clearness.

# References

Barton J (2006) Emerging patent issues in genomic diagnostics. Nat Biotechnol 24:939
Bera R (2009) The story of the Cohen–Boyer patents. Curr Sci 96:760

Berensmeier S (2006) Magnetic particles for the separation and purification of nucleic acids. Appl Microbiol Biotechnol 73:495

Goodmann L (1998) Random shotgun fire. Genome Res 8:567–568

Intellectual property rights and research tools in molecular biology (1997) Summary of a Workshop Held at the National Academy of Sciences, February 15–16, 1996. Washington, DC: The National Academies Press.

Jenei S (2005) Invitrogen patents not invalid where inventors did not appreciate their discovery. http://www.patentbaristas.com/archives/2005/11/20/invitrogen-patents-not-invalid-where-inventors-didnt-appreciate-their-discovery

Kling J (2005) Where the future went. In response to market trends and patenting laws, genomics companies are adapting their strategies EMBO reports 6, 1012–1014., doi:10.1038/sj.embor. 7400553

Ladanyi M (2005) Translating research into cancer molecular diagnostics and patents. J Clin Pathol 58:793

Little D (2006) FDA regulations and novel molecular diagnostic tests. CLI 7:48–49

Rajan M (2009) PCR (Polymerase Chain Reaction)—From 1983 to 2006—A chronological sequence. http://maliniqrs.blogspot.com/2009/11/pcr-plymerase-chain-reaction-from-1983. html

Saiki R, Scharf S, Faloona F, Mullis K, Horn G, Erlich H, Arnheim N (1985) Enzymatic amplification of beta-globin genomic sequences and restriction site analysis for diagnosis of sickle cell anemia. Science 20 230:1350–1354

Terry M (2006) Storming the molecular diagnostic IP fortress. Biotechnol Healthc J 49–54

Van Guilder et al (2008) Twenty-five years of quantitative PCR for gene expression analysis. Bio Techniques 44:619

# About the Authors

**Dr. Ulrich Storz** was born in 1969 in Muenster. He graduated in Biology from the University of Muenster in 1998, where he received his PhD in 2002. He is author and co-author of several scientific publications in the field of biology and biophysics as well as of several juridical publications in the field of intellectual property. He passed the German Patent Bar Examination in 2005. Since 2005, he has been admitted into practice as European Trademark Attorney at the European Trademark Office (OHIM). In 2006, he was registered in the list of representatives before the European Patent Office. This Main practice areas in the field of Intellectual Property Law include Patent Prosecution, FTO and Patent Infringement, as well as Patent strategies; especially in the Life Science field (i.e. Biotechnology, Biophysics, Biochemistry, microbiology). One of his major fields of interest is Antibody IP. Ulrich Storz is active as a speaker for the congress management company "Forum Institut für Management GmbH", and he organizes the annual Rhineland Biopatent Forum.

**Dr. Wolfgang Flasche** was born in 1977 in Marburg/Lahn. He graduated in Chemistry from the University of Marburg in 2002 after studying in Marburg and for nearly 2 years in Moscow at the MGU (Lomonosov Moscow State University). He received his PhD in Bioorganic Chemistry from the Humboldt University Berlin in 2006. He is author and co-author of several scientific publications in the field of chemistry and bio-organic chemistry. In 2011, he was registered in the list of representatives before the European Patent Office. As the Director IP for immatics biotechnologies GmbH he is responsible for the worldwide strategy and prosecution of all intellectual property.

**Kening Li** one of Wolfgang Flasches co-contributors, was born in China and obtained his BS in Biology from Nanjing University, and MS in Plant Pathology from Nanjing Agricultural University. He went to the U.S. in 1988, where he obtained his PhD in Molecular Biology in 1995, and J.D. in 1998, both from the University of Wisconsin—Madison. After law school, he practiced patent law with top-ranking U.S. law firms in Washington DC, as well in-house at Dupont Legal in

Wilmington, Delaware. He became a partner in the Washington DC office of Baker, Donelson, Bearman, Caldwell & Berkowitz, PC in 2006. He returned to China at the beginning of 2008, and now leads the Intellectual Property Team in China of Pinsent Masons LLP, a Global 100 law firm headquartered in London. For over a decade, Dr. Li has worked with clients ranging from start-ups to Fortune 100 companies, in managing IP assets as an integral part of their comprehensive competition strategy. He is experienced in formulating overall intellectual property strategy, resolving difficult freedom-to-operate obstacles, negotiating licensing agreements, and effectively utilizing existing intellectual property portfolio to strengthen clients' competitive positions. He is highly experienced in worldwide patent prosecution in the biotechnology, pharmaceutical and chemical arts, and in patent litigations in both China and the U.S., and Section 337 investigations before the U.S. International Trade Commission. Dr. Li is especially familiar with legal and commercial issues concerning technology transfer of academic or non-profit research organizations.

**Bryan Jones** the second co-contributor of Wolfgang Flasche, is an associate of Baker, Donelson, Bearman, Caldwell & Berkowitz, PC, Washington. Mr. Jones has advised clients regarding procurement and enforcement of patents and trademarks in the biotechnological, metallurgical, pharmaceutical, nutraceutical, and crop sciences. Mr. Jones also has experience in commercial and intellectual property litigation in state and federal courts. Prior to joining Baker Donelson, Mr. Jones was an extern for the Honorable Charles R. Norgle Sr. in the United States District Court for the Northern District of Illinois and was an intern with Liu, Shen and Associates, a private intellectual property firm in Beijing, China. As a law student, Mr. Jones served as an Associate Justice for the John Marshall Moot Court Executive Board, as a Staff Editor for the John Marshall Review of Intellectual Property Law, and as a competitor in the Jessup International Law and the Stetson International Environmental Law moot court competitions. Prior to law school, Mr. Jones worked as a research technician in Dr. Matthew Fenton's lab at Boston University School of Medicine and in Dr. William Klein's lab at Northwestern University.

**Dr. Johanna Driehaus** born in 1982 in Lüneburg, studied Biotechnology in Germany and Great Britain and graduated in 2005 from the University of Edinburgh. While working on her PhD thesis on human embryonic stem cells at the University of Bonn she first came into contact with the field of intellectual property. In 2011 she passed the German Patent Bar Examination. Her main practice areas include drafting and prosecuting patent applications, patent infringement and strategies for managing intellectual property rights including assessing the validity of these intellectual property rights in the areas of molecular diagnostics, biochemistry, biotechnology and microbiology.